**Rabaoui Ammar**

# Blockchain technology and online services

**Rabaoui Ammar**

# Blockchain technology and online services

## Digital technology and the decentralization of public action

**ScienciaScripts**

**Imprint**
Any brand names and product names mentioned in this book are subject to trademark, brand or patent protection and are trademarks or registered trademarks of their respective holders. The use of brand names, product names, common names, trade names, product descriptions etc. even without a particular marking in this work is in no way to be construed to mean that such names may be regarded as unrestricted in respect of trademark and brand protection legislation and could thus be used by anyone.

Cover image: www.ingimage.com

This book is a translation from the original published under ISBN 978-620-6-71843-7.

Publisher:
Sciencia Scripts
is a trademark of
Dodo Books Indian Ocean Ltd. and OmniScriptum S.R.L publishing group

120 High Road, East Finchley, London, N2 9ED, United Kingdom
Str. Armeneasca 28/1, office 1, Chisinau MD-2012, Republic of Moldova, Europe
Printed at: see last page
**ISBN: 978-620-7-99307-9**

# **Summary**


Blockchain is a digital solution for secure exchange, without intermediaries, between partners for all types of acts: transactions, smart contracts, voting, copyright, document storage... Thus, the research field of this work aims to define and analyze the possibilities of implementing Blockchain technology and its impacts on Tunisian online services and citizens' lives.

This technology has enormous development and application potential. Depending on the area in which it is implemented, Blockchain could solve some of the obstacles encountered and bring about changes in basic processes. Each of Tunisia's administrations, sooner or later, will have to face the changes caused by Blockchain. There is no technological or economic reason to force development, as this is a fundamental technology, not a disruptive one.

Consequently, there are many opportunities for decision-makers to use this work as an introductory point to a Blockchain-powered distributed future. This work can also serve as a starting point for further research, particularly in the Tunisian healthcare system, which is currently undergoing structural and organizational transformation. The healthcare ecosystem is seeing the emergence of a networked structure whose digital urbanization is expanding, and whose mastery of information management is bringing with it economic and legal challenges, as well as real prospects for the information professions.

**Keywords: Blockchain technology, online services, issues, information management, Tunisian healthcare system, development.**

# Thanks

The research work presented in this thesis was carried out at the Electronic Administration Unit of the Tunisian Government Presidency, in collaboration with the Ecole Nationale d'Administration de Tunis (ENA).

I would like to thank all the people who contributed to the completion of my thesis and helped me to write this report.

First of all, I'd like to thank Mr Khaled SELLAMI, Director General of the Electronic Administration Unit at the Presidency of the Government (until May 2023) for his support, advice and time.

I would like to thank all the members of the Electronic Administration Unit whom I came into contact with during the course of this thesis.

Finally, I would also like to thank all the people who kindly received me and discussed the various subjects of my report;

*Ammar Rabaaoui*
*Diploma in Applied Computer*
*Engineering, Master's degree in*
*Intelligent and Communicating*
*Systems (SIC) and post-graduate*
*diploma from the Ecole Nationale*
*d'Administration de Tunis.*

# Contents

# General introduction

Today, we are witnessing the simultaneous proliferation of Big Data, artificial intelligence, data science, robotics and other cutting-edge, rapidly emerging technologies. These cutting-edge technologies support and augment each other, affecting food systems, sanitation, energy, education, healthcare, social services and more.

Governments around the world, and in Tunisia in particular, understand the importance of positively supporting the emergence of these technologies and their uses. This is manifested in the form of reports, experiments in public use or legislative interventions designed to facilitate private achievements. It is in this context that Tunisia has introduced, for example, a law on the promotion of Startups[1] with the aim of encouraging young promoters to create economic value via NICTs.

In recent years, a major IT innovation known as 'Blockchain' has emerged as a potentially promising technology. Blockchain is a new concept that is driving and contributing to digital transformation in many sectors, particularly in the public sector. Several experiments have already been launched in a wide range of economic sectors. This prompts us to question the relevance of using Blockchain in relation to use cases, and the real motivations of the various players[2] .

This thesis aims to explore Blockchain and its uses. This technology, which promises substantial benefits for the sectors in which it will find application, is currently at an early, but rapidly evolving, stage of appropriation.

Research is essentially a proactive posture, most often expressed in the form of pilot applications designed to gain a better grasp of how the technology works, to prove that certain concepts of use are viable, to identify solutions to the contractual and regulatory problems posed, to

---

[1] "Draft law n ° 003/2018 concerning emerging companies".

[2] Berbain C., (2017), "La blockchain: concept, technologies, acteurs et usages", Série trimestrielle - Réalités Industrielles - Blockchains and smart contracts: technologies of trust? , page 6

build new business relationships and to learn how to accompany these changes.

However, the major changes that Blockchain's uses could introduce can only encourage public administration players to take an interest in this technology and develop innovative projects that will enable them to keep up with or even stay ahead of constantly evolving administrative practices.

This dissertation aims to outline a broad path for public intervention in the appropriation of Blockchain technology. This path presents the State as a regulator of the uses of this technology by initiating a reflection on the definition of the rules of the game enabling the protection of citizen-consumers or by setting the quality rules that Blockchain technology must respect to be considered reliable with regard to a use, to guarantee the execution of legal decisions via this technology or to apply the rules of respect for privacy or the fight against fraud.

As in the case of digital transactions, questions will also arise concerning the territoriality of operations carried out using blockchain technology. As a result, strengthening the security of this technology could also involve consideration of the opportunity to develop national capabilities or to support private players in their development of applications including this technology.

In addition, the progressive growth of data generated by our exchanges, which is one of the dimensions of Big Data, raises issues of authenticity control, controlling the risks of piracy and sharing individual data, to which Blockchain technology is trying to provide a solution.

As a result, the essential idea behind Blockchain technology is to create a distributed ledger that operates without a centralized control body and can be updated in real time by all parties involved in an exchange, by combining it with classic cryptographic methods that ensure control over the level of information shared.

Blockchain, this new technology, has an impressively versatile potential. After its initial use in the field of virtual currencies known as

Bitcoin, it is now beginning to be tested in fields as varied as energy, transport and automobiles[3] ....

In order to grasp this new concept, we need to define it, identify its structuring elements and question the relevance of its properties and promises.

Technically, Blockchain technology can be seen as the combination of three technologies[4] : a peer-to-peer sharing system on a network, thus constituting the shared ledger, algorithms for validating new entries written to this ledger, and advanced cryptographic techniques for securing data exchanges or transactions. This combination is the origin of an infrastructure generating trust in the realization of value exchanges between stakeholders without the need for an intermediary.

Blockchain technology can therefore potentially replace all cases where such a central authority has no other use than to act as an intermediary. As a result, this technology can be seen as a challenge to centralized powers, since it is the first time in the history of technological revolutions that a technology has the capacity to act on the vertical and centralized power exercised by states **in very diverse fields ranging from corporate auditing, to electoral and voting systems in general, land ownership management and** real estate transfers, **insurance systems,** electricity or fuel distribution....

However, the move towards the use of Blockchain, may be stymied by three factors of uncertainty: the semi-mature evolution of this technology as well as the underlying algorithms, the calibration of the profitability feedback of investments in this field over a long period, and the current non-formalization of regulation. Nevertheless, and in view of its potential impact and despite the uncertainties, it is essential that public or private decision-makers acculturate themselves on Blockchain technology and study its possible implications on their activities, in

---

[3] MEDEF, (2016), "La blockchain pour les entreprises - Soyez curieux! Understand and experiment", page 9

[4] MEDEF, (2016), "La blockchain pour les entreprises - Soyez curieux! Understand and experiment", page 11

particular, guaranteeing the interoperability, transparency and confidentiality of systems and networks under all conditions.

As such, this thesis aims to explain the concepts behind Blockchain technology, take stock of the progress of this revolution and identify a few key questions that economic players and decision-makers will need to ask themselves to take advantage of this revolution.

More specifically, this dissertation highlights, in its first part, how Blockchain technology works by answering the following question: what does the concept of "Blockchain technology" cover and what are its functionalities?

The second part of the dissertation sets out to explore the possible uses of Blockchain technology in a Tunisian context, particularly in public activities, and to demonstrate the interest for decision-makers in experimenting with this technology. Thus, this section addresses the following problematic: What is the position of national public administrations in relation to the use of Blockchain technology? And are there any specific applications worth exploring in case studies?

Finally, the third part will be devoted to proposing a possible integration of this technology within the healthcare sector, with the aim of promoting healthcare services for the benefit of Tunisian citizens and strengthening the resilience and sustainability of Tunisia's healthcare systems. This involves answering the question: What is the most appropriate technological approach to ensure this integration for future use in this sector?

# Chapitre I.   Concept and issues

## Introduction

In this chapter, we will try to present the evolution of digital governance and the interest of approaching the concept of Blockchain (1). We will then describe this technology (2) and its origins (3). Next, we'll look at its technical aspects in terms of structure (4), types (5) and properties (6). Then, we'll illustrate the usefulness and potential of Blockchain technology (7), as well as the opportunities and challenges of implementing such a technology (8).

## 1. Evolution of global digital governance

E-government has become an integral part of the transformation of the public sector.  Indeed, information and communication technologies (ICTs) have made it possible to provide more modern services to citizens and businesses, stimulate the information society and new emerging economies, and boost the transformation of the public sector.

Innovations such as Cloud Computing and Big Data, alongside the rapid spread of the internet, have fostered the emergence of digital platforms as the preferred channel for service delivery.  Governments and the way they deliver services to their citizens are following similar patterns.

In addition, since early 2003, the United Nations has regularly published its E-Governance Survey, which assesses the development of national e-government capabilities in the 193 member states of the United Nations. The proposed global ranking is based on an indicator called EGDI[5] (E-Governement Development Index), which covers three key indices: online services, telecommunications infrastructure and human capital. The three indices that make up the EGDI cover a wide range of topics relevant to e-government:

---

[5] "UN E-Government knowledgebase", available at https://publicadministration.un.org/egovkb/en-us/ ;

- The Online Service Index measures a government's ability and willingness to provide services and communicate with its citizens electronically.
- The telecoms infrastructure index measures the existing infrastructure required for citizens to participate in e-government.
- The human capital index is used to measure citizens' ability to use e-government services.

According to the latest United Nations "United Nations E-Government Survey 2018" report[6] , and as in 2016, Tunisia is still positioned in the group of African countries with a high level of EGDI. Ranked 108ème in the first edition of the report in 2003, 103ème in 2012, Tunisia is currently ranked 80ème worldwide in the "e-government" ranking and 3ème in Africa after the Republic of Mauritius and South Africa.

Today, governments and citizens around the world coexist in a digital governance ecosystem facilitated by a wide range of interconnected technological factors providing services to citizens in areas such as health, identity and social services, etc...

The emergence of new technologies, coupled with changes in public behavior, explain the huge success of e-government. The future of e-administration is promising, given the positive response from users and the appreciation of its benefits by civil servants. Many projects to improve public services are still under study, with the aim of making daily life easier for citizens. The reality is that digitization will affect all administrative services, enabling all procedures to be carried out in their entirety.

However, the projection of e-government must be accompanied by balanced and appropriate governance. The growing demand from users to be at the heart of the administration's concerns calls for different and truly innovative methods in the relationship between governors and governed. The development of digital services should provide the best possible

---

[6] United Nations Department of Economic and Social Affairs, (2018), "United Nations E-Government Survey 2018", Copyright © United Nations, 300 p.

response to the challenge of modernizing government by encouraging new uses. It must also enable us to meet the political and democratic challenge.

By adopting a fully participative approach, digital technology should enable us to enhance the value of individuals, revitalize relations between the various players and open up the field of creativity in the service of the general interest.

Thus, the combined use of creative innovations in the operation of public services is a tool for innovation in government. Blockchain technology represents one of the technological facets of this innovation. The stakes of using this technology are major, and promote the notion of sharing in its economic and social, political and legal dimensions[7].

The enthusiasm generated by Blockchain technology and its potential applications have led to the rapid formation of a rich ecosystem in which several types of players, with varying motivations, are involved:

- Many start-ups have launched, including those that provide advice;
- Major groups, particularly in the financial sector (banks, insurers, etc.), are experimenting with this approach in partnership with start-ups or with research and innovation laboratories or units;
- Public authorities, both in Tunisia and abroad (the Tunisian Post Office, France Stratégie[8], UK Government Office for Science[9]) are accompanying or observing the phenomenon.

As a result, a key question for governments, and Tunisia in particular, is now how to adopt this digital innovation in their policies and find ways to implement it in their role as facilitator and service provider. The rest of this paper will attempt to answer this question.

## 2. Definition and operating mode of Blockchain technology

---

[7] Faridah D., and Gallouj F, (2012) "Innovation in public services", Revue française d'économie, volume xxvii, no. 2, pp. 97-142 ; Available at https://www.cairn.info/revue-francaise-d-economie-2012-2-page-97.htm

[8] "Le volet "Blockchains" de France Stratégie", available at http://www.strategie.gouv.fr/chantiers/blockchain ;

[9] "Distributed ledger technology: beyond block chain", available at https://www.gov.uk/government/news/distributed-ledger-technology-beyond-block-chain ;

Blockchain is a technology for storing and transmitting information that is transparent, secure, and operates without a central control body (Blockchain France definition)[10] .

Blockchain is a computer process that enables 100% authentication of any transaction, such as the exchange of money or the transfer of confidential information. Taking the example of an online purchase, by using Blockchain technology, we no longer need to go through a financial intermediary who certifies that we have money in our account. This allows third parties who don't know each other to make secure transactions without an intermediary. The question is, how does it work?

Referring to mathematician Jean-Paul Delahaye, a Blockchain can be likened to "a very large notebook, which anyone can read freely and for free, on which anyone can write, but which is impossible to erase and indestructible."[11]

Blockchain technology takes the form of a large virtual register in which all transactions are stored. The data is encrypted, and each time a transaction is carried out, a small piece of information called a "block" is created (as shown in figure 1 below). Blockchain is therefore a succession of interconnected mathematical blocks containing the history of all exchanges carried out between its users since its creation.

---

[10] Blockchain France, (2016), "La Blockchain décryptée-Les clefs d'une révolution", Observatoire Netexplo © Blockchain France Associés, page 1

[11] Blockchain France, (2016), "La Blockchain décryptée-Les clefs d'une révolution", Observatoire Netexplo © Blockchain France Associés, page 2

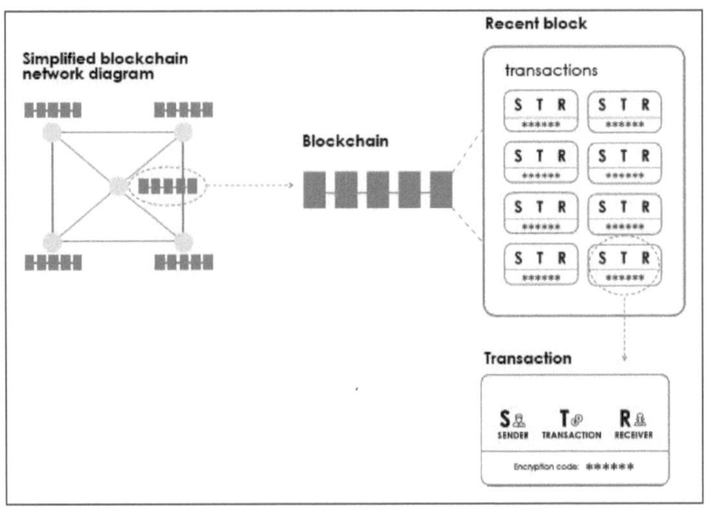

**Figure 1Simplified illustration of a distributed ledger network (Blockchain)[12]**

The following figure (Figure 2) shows how Blockchains reach agreement.

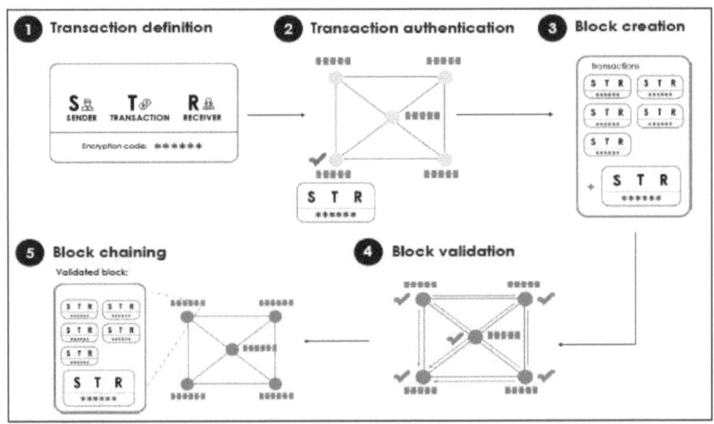




[12] EVRY, (2016), "Blockchain: Powering the Internet of Value", Whitepaper @ evry's labs, page 9; available at https://www.evry.com/en/news/articles/banking-on-the-blockchain/

**Figure 2Overview of a Blockchain transaction**[13]

Each node in the decentralized Blockchain network contains a copy of a transactional record: data modified in a node must be authenticated by peer approvals once it is replicated on the network. Blocks, built up from a growing list of records as a basic foundation, are linked and secured by a strong public-key cryptographic infrastructure. These blocks are linked by a hash pointer (cryptographic hash function), which guarantees that transactions are free from forgery.

Each block contains a set of transactions, the size of which depends on the number of transactions carried out. The execution time of the process varies according to the blockchain in question. But where does this technology come from?

## 3. Origin and development of blockchain technology

To understand more about Blockchain, we need to go back to the source. Indeed, any integration or development of a new technology or online service essentially relies on the degree of trust of the various users. In this sense, two major innovations have revolutionized the way trust is generated: these are advances in cryptography and distributed computing architectures. The following Table 1 illustrates a brief history of this evolution, which led to the birth of Blockchain technology.

| Date | Evolution |
| --- | --- |
| **1976**<br>**Cryptography asymmetrical**<br>**:**<br>**double**<br>**concept**<br>**public key /**<br>**private** | In 1976, American researchers Whitfield Diffie and Martin Hellman introduced the concept of dual public and private keys.<br><br>The exchange between two agents must be encrypted without the need for a password, thanks to the Diffie-Hellman protocol. This innovation is the genesis of blockchain technology.<br><br>This theoretical innovation has been accompanied by an exponential increase in computing power associated with the availability of computing units, hence the emergence of the notion of Cloud computing. |
| **1990**<br>**The**<br>**architectures** | Distributed architectures are setting the standard for stability and security. One example is the Web (HTML), which was born in the CERN laboratories in the early 1990s. |

---

[13] EVRY, (2016), "Blockchain: Powering the Internet of Value", Whitepaper @ evry's labs, page 10; available at https://www.evry.com/en/news/articles/banking-on-the-blockchain/

| distributed : the Web | In addition, we can mention domain name management and the associated domain name servers (DNS), distributed and replicated in the Internet's "nodes".<br><br>All these examples have never failed since their creation. |
|---|---|
| | ➤ Cryptography and trust-generating distributed architectures have converged to form Bitcoin's technological layer: the Blockchain. |
| 2008<br>**Birth of the Blockchain** | In 2008, Satoshi Nakamoto[14] presented a method for solving a cryptographic problem: the double payment or Byzantine Generals problem. This prevented two agents from exchanging assets without going through a trusted third party.<br><br>The solution is based on the decentralized architecture that supports Bitcoin: the blockchain.<br><br>Through this discovery, two agents who don't know each other can exchange assets without the transaction having to be secured and validated by a central authority. Thus, the trust intrinsically created by the Blockchain is a disintermediation tool whose direct effects are to reduce costs and make exchanges more fluid. |
| | ➤ Payments, transactions, contracts: blockchain technology is the possible catalyst for a totally disruptive approach. |

**Table 1:Historical development behind blockchain[15]**

Over the past decade, Blockchain technology has been developed as the main component of Bitcoin, the leading cryptocurrency. Since then, several Blockchains have emerged as public registries for specific types of transaction. For example, companies like IBM and Samsung are using Blockchain technology for a network of decentralized Internet of Things[16] (IoT) devices, operating as a public ledger to enable them to communicate directly with each other to update software, manage bugs and monitor power consumption.

In this way, the Blockchain has evolved from a mechanism for powering the Bitcoin crypto-currency to a technology ready to be

---

[14] Nakamoto S., (2008), "Bitcoin: A Peer-to-Peer Electronic Cash System", 9 p.; available at www.bitcoin.org

[15] Guillaume B., (2016), "Understanding the Blockchain - anticipating the disruptive potential of the Blockchain on organizations", uchange.co, page 8

[16] "IBM Reveals Proof of Concept for Blockchain-Powered Internet of Things", available at https://www.coindesk.com/ibm-reveals-proof-concept-blockchain-powered-internet-things/;

integrated into a variety of fields. Its ongoing development can be divided into four stages[17] :

- **Blockchain V 1.0:** this stage is characterized by applications in cryptocurrencies, particularly Bitcoin, thanks to the implementation of distributed ledger technology.

- **Blockchain V 2.0:** this stage recognizes the emergence of smart contracts, to facilitate, verify and execute transactions. The emergence of the Etherium ecosystem is a case in point, as is the evolution of decentralized applications using decentralized communication storage protocols.

- **Blockchain V 3.0:** this is the stage of development and maturation of authorized Blockchains, with innovations in throughput and confidentiality adapted to the enterprise.

- **Blockchain V 4.0:** this is the next stage in the evolution of Blockchain and is characterized by industry-wide deployment of value chains, including enterprise systems integration and borderless process automation; these developments should ultimately increase transparency and trust between companies, employees and customers.

The original Bitcoin network[18] , which appeared in 2009, was built to secure the Bitcoin crypto-currency. It has around 5,000 complete nodes and is distributed worldwide. It is primarily used to exchange Bitcoin and trade value, but the community has seen the potential to do much more with the network. Because of its size and proven security, it is also used to secure other Blockchain applications.

[17] Mishra R. et al, (2018), "How Integrated Process Management Completes the Blockchain Jigsaw," Digital Systems & Technology @Cognizant, page 7
[18] Delahaye J., (2016), "Cryptographic currencies & blockchains", Université de Lille 1@ INRIA Saclay; available at http://cristal.univ-lille.fr/~jdelahay/LeBitcoin/DelahayeBitcoin23-6-2016.pdf

The Ethereum network[19] , launched at the end of 2013, is a second evolution of the Blockchain concept. It takes the traditional Blockchain structure and adds a programming language that is built inside it. Like Bitcoin, it has over 5,000 complete nodes and is globally distributed. Ethereum is mainly used to exchange "ether" (a currency used as a financial intermediary to run smart contracts), establish smart contracts and create decentralized autonomous organizations (DAOs). It is also used to secure various blockchain-based applications.

The Factom network[20] , launched in September 2015, is a third evolution of Blockchain technology. It uses a lighter consensus system, incorporates voting and stores much more information. It was built primarily to secure data and the system. Factom works with federated nodes and an unlimited number of auditing nodes. Its network is small, so it attaches to other distributed networks by building bridges across blockchain entries.

Of course, Blockchain technology ultimately enables the creation of a decentralized system, as information is distributed across the computers of connected people, like peer-to-peer. This system ensures traceability, trust, decentralization and disintermediation. So, what's the point of this type of system?

## 4. The structure of blockchains

Blockchains are made up of three main parts[21] : the block, the chain and the network.

### ❖ Block

This is a list of transactions recorded in a ledger over a given period. The block size, period and triggering event are different for each blockchain.

---

[19] Ethereum community, (2017), "Ethereum Homestead Documentation, Release 0.1", Ethereum Homestead Documentation , page 8

[20] "Factom", available at https://www.okchanger.fr/cryptocurrencies/factom ; **And** "Factom Foundation", available at " https://www.factom.com/ ;

[21] Laurence T., (2017), "Blockchain for Dummies", John Wiley & Sons, Inc, Hoboken, New Jersey, 214 pp.

Not all Blockchains are recording and securing a record of the movement of their crypto-currency as their primary objective. But all Blockchains must enable this type of recording. You have to think of the transaction as simply the recording of data. Thus, the assignment of a value (as in a financial transaction) is used to interpret what this data means.

## ❖ Chain

A chain is a hash that links one block to another, chaining them together mathematically. This is one of the most difficult concepts to understand in Blockchain technology. It's this linking that enables the various Blockchains to create mathematical trust.

The hash in the blockchain is created from the data in the previous block. The hash is an imprint of this data and locks the blocks in order and time.

Although blockchains are a relatively new innovation, hashing is not. Hashing was invented over 30 years ago. This old innovation is used because it creates a one-way function that cannot be decrypted. A hash function creates a mathematical algorithm that maps data of any size into a fixed-size bit string. A bit string is typically 32 characters long, representing the data that has been hashed. The Secure Hash Algorithm (SHA) is one of the cryptographic hash functions used in blockchains. SHA-256 is a common algorithm that generates an almost unique 256-bit (32-byte) hash of fixed size.

## ❖ Network

The network is made up of nodes known as "complete nodes". They gather at a computer running an algorithm that secures the network. Each node contains a complete record of all transactions that have been registered in a blockchain. Nodes are located anywhere in the world and can be operated by anyone. It's difficult, expensive and time-consuming to operate a complete node, so people don't do it for free. They are incentivized to mine a node because they want to earn crypto-currency. The underlying Blockchain algorithm rewards them for their service. The reward is usually a token or a crypto-currency, such as Bitcoin.

It should be noted that the terms Bitcoin and Blockchain are often used interchangeably, but they are not identical. Indeed, Bitcoin has a Blockchain that functions as an underlying protocol enabling the secure transfer of Bitcoin, and the term Bitcoin is the name of the crypto-currency that powers the Bitcoin network. So, the Blockchain is a class of software, and Bitcoin is a specific crypto-currency.

## 5. Blockchain type

Blockchains can be thought of as distributed databases controlled by a group of individuals, which store and share information.

A Blockchain is a data structure that enables a digital record of data to be created and shared between a network of independent parties. There are many different types of Blockchain[22] :

❖ **Public blockchains:** public blockchains, such as Bitcoin, are large distributed networks that are run via a native token. They are open to anyone to participate at any level and have open-source code that their community maintains.

❖ **Authorized blockchains:** authorized blockchains, such as Ripple, control the roles that individuals can play in the network. They are always extended, distributed systems that use a native token. Their code base may or may not be open source.

❖ **Private blockchains:** private blockchains tend to be smaller and do not use tokens. Their membership is tightly controlled. These types of Blockchains are favored by unions that have trusted members and exchange confidential information.

All three types of Blockchain use cryptography to enable every participant on a given network to manage the ledger securely without the need for a central authority to enforce the rules. Removing the central authority from the database structure is one of the most important and powerful aspects of Blockchains.

Blockchains create permanent records and transaction histories, but nothing is truly permanent. The permanence of the record is based on the

---

[22] Laurence T., (2017), "Blockchain for Dummies", John Wiley & Sons, Inc, Hoboken, New Jersey, 214 pp.

permanence of the network. In the context of Blockchains, this means that a large part of a Blockchain community would have to agree to change the information and have an incentive not to change the data.

When data is stored on a blockchain, it is extremely difficult to change or delete it. When someone wishes to add a record to a blockchain, also known as a transaction or entry, network users with validation control check the proposed transaction. This is where things get tricky, because in every blockchain, you have to find out how it should work and who can validate a transaction.

## 6. Blockchain technology properties

Given that blockchain technology brings together three technologies - decentralized architecture, cryptographic protection and crypto-currency or transaction issuance - it implies the integration of three properties[23] in relation to these technologies, namely disintermediation, security and autonomy.

### 6.1. Disintermediation

The first property of Blockchain technology is to provide the trust necessary for users to exchange information without the control of a trusted third party. Consensus thus replaces centralized validation.

Blockchains are powerful tools, because they create honest, self-correcting systems without the need for a third party to enforce the rules. They accomplish rule enforcement through their consensus algorithm.

In the world of Blockchain technology, consensus is the process of developing an agreement between a group of generally distrustful shareholders. These are the complete nodes on the network. Complete nodes validate transactions entered into the network to be recorded in a database, thus forming a distributed ledger.

Each blockchain has its own algorithms for creating agreement within its network on added entries. There are many different models for creating

---

[23] Guillaume B., (2016), "Understanding Blockchain - anticipating Blockchain's disruptive potential on organizations", uchange.co, pages 11-15

consensus, as each Blockchain creates different types of input. Some blockchains are traded, others store data and others secure systems and contracts.

In the banking system, for example, international transfers are costly and take several days to process. In contrast, a transfer using a crypto-currency like Bitcoin is virtually instantaneous, secure and free of charge.

But for a transaction to take place on a Blockchain network, the information linked to it must be integrated into a block. To achieve this, the transaction must be validated by several network nodes, known as miners, who verify its conformity by solving a complex cryptographic problem. This result can be verified collectively using the "Proof of Work" protocol[24] . The whole operation is called "mining"[25] . Once all the miners agree on the validity of the "Proof of Work", the transaction is integrated into a block, which is then added to the "block chain".

The mechanism for adding new blocks is the result of a consensus between network players, making control by a reference institution unnecessary. This consensus is the vector of disintermediation, and is reflected in the collective validation[26] of the "Proof of Work".

### 6.2.Security

There are two mechanisms that guarantee the structural security of information stored within a system based on Blockchain technology. The first is the cryptographic process: the code of each new block is built on that of the block preceding it in the blockchain, making it impossible for a change in one block to change all the blocks in the chain. The second mechanism is the decentralized architecture: within a Blockchain architecture, all the blocks are replicated in the nodes of the network, and not in a single server. This decentralized architecture acts as a structural defense against the risk of data theft.

---

[24] Morabito V., (2017), "Business Innovation Through Blockchain, The B³ Perspective," © Springer International Publishing, page 10

[25] Morabito V., (2017), "Business Innovation Through Blockchain, The B³ Perspective," © Springer International Publishing, page 70

[26] Morabito V., (2017), "Business Innovation Through Blockchain, The B³ Perspective," © Springer International Publishing, page 69

As a powerful method of securing digital transactions via this distributed cryptography infrastructure, Blockchain has become the new "trusted protocol". This technology is now seen as having the potential to solve the double-spending problem (also known as the 51% attack[27] ) that plagues many industries, without the need for a trusted authority or central server.

Indeed, the 51% attack is one in which a hacker manages to possess at least 50% of the mining computing power. As a result, he or she will theoretically be able to create blocks faster than any other miner, and subsequently create a forged blockchain alongside the existing one. So, the tamper-proof design of the blockchain is the key to its security.

### 6.3. Autonomy

Today, online services are backed by platforms that take care of the infrastructure requirements. In the case of Blockchain technology, computing power, measured in hash/second, and hosting space are provided by the network nodes themselves. Consequently, within a Blockchain, the infrastructure is no longer concentrated at the level of one organization, but is, on the contrary, spread across all points of the network. In this way, a Blockchain is self-supporting and independent of third-party services.

## 7. Utility and potential of blockchain technology

Blockchain is a fast-growing technology with the potential to revolutionize the way information is handled and exchanged. This can be explained by enumerating the strengths of this technology, illustrating potential for change in sectoral examples and citing some international experiences.

### 7.1. The strengths of blockchain technology

The main strengths of Blockchain technology lie in building trust and integrating sound principles to establish a Blockchain-based economy.

#### 7.1.1. Building trust

---

[27] "What is a 51% or double spending attack?", available at https://cryptoast.fr/quest-ce-quune-attaque-a-51-double-spending/ ;

Blockchains are now recognized as the "fourth evolution" of computing, the missing layer of trust for the Internet. This is one of the reasons why so many people have become excited about the subject.

Ever since the advent of the Internet, privacy and security have become major issues. In the 1980s[28] , engineers tried to address these issues using cryptography, but to no avail. At the time of the 2008 financial crash[29] , Bitcoin, a new crypto-currency, appeared on the scene. This peer-to-peer exchange protocol, which is not controlled by any country or entity and whose code is free, has the ability to ensure the integrity of exchanged data, without going through a third party (bank or other).

Indeed, Blockchain's operation relies on the willingness of individuals to make their computers available to host part of the protocol. So the notion of a central server that could be hacked does not exist. What's more, the Blockchain is encrypted, since it relies on a set of public and private access keys.

Thus, Blockchains can create trust in digital data because once information has been written to a Blockchain database, it is almost impossible to delete or modify it. This capability has never existed before.

Every 10 minutes[30] , all transactions are verified and stored in a block that relates to the previous one, thus creating a chain. So, to steal a Bitcoin, you'd have to retrace and rewrite its entire history on the Blockchain, in full view of everyone.

Potentially, with Blockchain, trust arises from the community, not from third parties certifying the transaction. Thus, four principles must be respected to create trust in a business environment: honesty, empathy, a sense of responsibility and transparency. Today, the Blockchain makes it possible to envisage a new conception of identity: rather than having a multitude of documents (driving license, birth certificate...), we would

---

[28] Tapscott D. and Tapscott A., (2016), "Blockchain Revolution," Penguin Random House LLC (version1), page 30

[29]  Tapscott D. and Tapscott A., (2016), "Blockchain Revolution", Penguin Random House LLC (version 1), page 30 **And** Magnen J. and Fourel C., (2015), "Mission d'étude sur les monnaies locales complémentaires et les systèmes d'échange locaux", report, Part One, sec.secacess-presse@cabinets.finances.gouv.fr, page 11

[30] Tapscott D. and Tapscott A., (2016), "Blockchain Revolution," Penguin Random House LLC (version 1), page 31

only have one identity inscribed in the Blockchain that would deliver the desired information at the desired moment.

### 7.1.2. Solid principles for a blockchain-based economy

Since 2008, numerous projects based on Blockchain technology have emerged. Each of them is based on implicit principles that shed light on what Blockchain technology is. There are seven principles for imagining a Blockchain-based economy[31] :

- **Integrity:** in the blockchain, the network is the guarantor of its own integrity, as the system is based on transparent exchanges. For example, in the case of Bitcoin, the system marks Bitcoin spent and prevents it from being transferred elsewhere, thus preventing fraud. Each transaction is verified by the chain's participants, known as "miners", until a consensus is reached on its validity. To reach this consensus, the miners have to work hard to decipher a mathematical puzzle that guarantees to all the others that the job has been well done, and that the analyzed block can be added. As a result, users can trust the network rather than large corporations or governments.

- **Distributed verification:** at the blockchain level, verification of transactions is shared between different members of the network, decentralizing power and making it impossible to attack the entire system. As a result, no bank account can be frozen or seized.

- **Cooperative participation:** the third principle of Blockchain is that value becomes an incentive to make the system work. With Blockchain, the community has an interest in the system's functioning. By serving their own interests, users are therefore working for the whole community.

- **Security:** security is at the heart of network operation, since it guarantees the confidentiality and authenticity of activities. Logically, the longer the data chain, the more secure it is, since it is more difficult to trace it back to its source and try to falsify it.

---

[31] Tapscott D. and Tapscott A., (2016), "Blockchain Revolution," Penguin Random House LLC (version 1), pages 47- 66

- **Data control:** it's obvious that people need to be able to control their data, and that they have a right to privacy. Blockchain technology offers the opportunity to escape a surveillance society where all our data escapes us.

- **Transparency:** in a solution based on Blockchain technology, property rights are transparent and respected: an element is only transmitted if it belongs to its issuer, and the latter is the sole decision-maker on data issuance.

- **Inclusion:** the final principle of Blockchain technology is inclusion, which means that everyone should benefit from the way the system works. Even today, a significant proportion of the population does not have access to information and communication technologies. Blockchain aims to change this, so that a simple telephone can be used to participate in the exchange of information: a micropayment service, for example, enables users to top up their phone credit by texting an amount in Bitcoin.

## 7.2. Potential mobilized in sectoral examples

Several sectors and fields can be affected by an innovative management model through Blockchain technology. For example, modernization can affect financial services, business organization methods and the way public actions and decisions are governed.

### 7.2.1. Reinventing financial services

Whereas yesterday's financial services were opaque, slow and hierarchical, the new order promises to be more transparent, egalitarian, secure and private. One of the major problems with finance is that it is a monopoly in its own right, enabling the banks to set very high transaction costs. They therefore have no reason to want to change or improve the system. With Blockchain technology, we can renovate finance by putting an end to the sector's monopoly situation.

Blockchain technology enables two people wishing to enter into a transaction to do without a trusted third party to verify the other's identity, while at the same time lowering the transaction costs borne by traditional banking and increasing the speed of exchanges.

Several properties of this technology could revolutionize the financial sector. The ability to make secure identifications, transfer, store, lend, exchange values, be able to invest, insure risks and facilitate controls make the Blockchain a fast and trusted system.

While banks based their model on asymmetric information vis-à-vis their customers, Blockchain technology was built on transparency. Private Blockchains have therefore been developed. These are systems where only certain users have access to all information. The problem with these restricted Blockchains is that one member can more easily change the rules without the community's agreement.

It is therefore necessary to define common standards for this new technology. This raises the question of the role of governments. Should governments intervene in the regulation of such an industry? At the moment, innovation is moving at a pace they can't keep up with, yet they can't afford to do without supervision.

What's more, modern accounting faces a number of problems that this technology can solve. On the one hand, managers always swear that their accounts are accurate... In reality, however, they are sometimes wrong. Blockchain technology could prove managers' wilful errors. On the other hand, Blockchain technology could make up for human error, which is omnipresent in accounting problems.

Finally, traditional accounting methods are not adapted to new models such as micro-transactions. The problem lies not in accounting itself, but in the way it is set up in companies. In addition to the debit and credit columns on their balance sheets, companies could add a balance sheet, which would record transactions in real time. Afterwards, they could give access to their data to selected persons such as shareholders, auditors.... Simplified access for greater system integrity.

At present, bank customers are given a score to enable them to take out a loan. This score reflects the trust that the bank places in its customer, but it often obeys very restrictive criteria that often do not reflect a real ability to repay. Blockchain technology would make it possible to establish a new standard of trust based on a customer's reputation and involvement.

### 7.2.2. The evolution of companies towards new forms of organization

Tomorrow's leaders will have the opportunity, thanks to Blockchain technology, to change the value creation process, negotiating without constraint with their suppliers, customers, etc. But, despite new management styles and the desire to create horizontal hierarchies, the new digital companies have all adopted top-down hierarchies. So what's the point of using this technology for companies that make money by collecting data? Following the theory of British economist Ronald Coase[32] , the use of such technology would lower internal and external transaction costs. It is this gain that could lead companies to profoundly alter their organizational structures.

Blockchain technology can also revolutionize the way contracts are drawn up, where the costs involved in negotiating them are not negligible. For example, this technology makes it possible to set up "smart contracts", which activate when signatories fulfill contractual conditions.

Similarly, the status of the contract could be easily consulted, to see whether the goods have been dispatched, for example. Contracts involving several stakeholders can also be negotiated. Each of the members would then be given a specific key, which they would activate once the transaction has been completed.

Thus, through Blockchain technology, the redistribution of power and authority, and project-based working, promote innovation, productivity and the quality of customer relations. Indeed, Blockchain technology makes it possible to do business even when trust is lacking between two parties, as the integrity of all is ensured by the system. As a result, companies will have to redefine their fields of action, to focus on their core business.

### 7.2.3. New ways of governance and democracy

Blockchain technology can improve the quality of service and efficiency of governments by promoting their integrity and transparency. All personal data would be stored on a Blockchain instead of being centralized in different government agencies. Everyone would then be the

---

[32] Tapscott D. and Tapscott A., (2016), "Blockchain Revolution," Penguin Random House LLC (version 1), page 139

full owner of their data. In Estonia, for example, citizens each have a digital identity that enables them to carry out administrative procedures online, and even to vote.

This technology can also give some decision-making power back to local communities. Transparency is increased, as facts are easily verifiable. What's more, by making certain data public, we increase the usefulness and trust we place in them. The use of smart contracts for our political representatives can help monitor their power to act and decide, as well as ensuring that they live up to their campaign commitments.

However, we shouldn't try to replace representative democracy with technology, but it can help to glimpse the outlines of a new model. Indeed, Blockchain technology could make it possible to vote securely and indisputably by creating a network where voters attribute their votes to the account of the candidate of their choice. This means that votes can no longer be manipulated or falsified.

Blockchain technology could also enable the invention of new procedures. For example, disputes between private individuals could be resolved via the Blockchain, thanks to the evidence gathered on the network, which can be confronted with the terms of the smart contract. Similarly, the technology could revolutionize the judicial system by combining transparency and online citizen participation. So, thanks to its characteristics, Blockchain technology can guarantee the development of several new democratic tools.

### 7.3. Use cases for blockchain technology

Hundreds of Blockchain applications exist today. The whole world has become obsessed with the idea of transferring money faster, embedding and directing it in a distributed network, and building secure applications and hardware.

Most Blockchain applications handle the transfer of money or other forms of value quickly and inexpensively. This includes trading shares in a public company, paying employees in other countries and exchanging one currency for another.

Blockchains are also being used as part of a software security stack. The US Department of Homeland Security[33] is studying blockchain software to secure Internet of Things (IoT) devices. The world of the Internet of Things offers the most advantages, as it is particularly vulnerable to identity theft and other forms of hacking. IoT devices have also become more widespread, and security has become more dependent on them. Hospital systems, autonomous cars and security systems are perfect examples.

DAOs or "decentralized autonomous organizations" are another interesting innovation in Blockchain technology. This type of Blockchain application represents a new way of organizing and incorporating online businesses. DAOs have been used to organize and invest funds via the Ethereum network.

Larger-scale projects being explored via Blockchain technology now include government-backed land registration systems, identity and international travel security applications.

Countries[34] like the UK, Singapore and the UAE see it as a way to cut costs, create new financial instruments and keep clean records. They have active investments and initiatives exploring blockchain.

The social and economic implications of Blockchain applications can be emotionally and politically polarizing, as Blockchain technology will change the way we structure value-based and society-based transactions.

Blockchain technology[35] has come a long way since its public awareness due to a significant number of news reports in 2015. Since then, many startups have been working on beta and pre-launch versions, with nearly 2,000 new Blockchain startups forming in 2016. Many of these are finally being commercialized in 2017 and will be again in 2018 in

---

[33]  Labrot É. and Ségur P., (2011), "Un monde sous surveillance?", © Presses universitaires de Perpignan, pages 227-241 ;

[34] Laurence T., (2017), "Blockchain for Dummies", John Wiley & Sons, Inc, Hoboken, New Jersey, pages 164-166

[35] Laurence T., (2017), "Blockchain for Dummies", John Wiley & Sons, Inc, Hoboken, New Jersey, page 163

Singapore, Dubai and London, where regulators are hailing the innovation and vying to become the world's financial powerhouse.

### 7.3.1. Progress in the United Kingdom

In 2016, the UK central government published a report entitled "Distributed Ledger Technology: beyond block chain"[36] , which claimed that distributed ledger technology (Blockchains) could be used to reduce corruption, errors and fraud, and make various processes more efficient.

They also said that Blockchain technology could change citizens' relationship with their government by bringing more transparency and reliability. But London has been very friendly to the technology since at least 2014. Many Blockchain start-ups incorporated or worked in London because it was the unofficially safest place to establish business. It was a big deal back then because a lot of crypto-currency entrepreneurs were arrested in 2014 and 2015.

Since that report was published, Blockchains have been approved for government applications in the UK, including Whitehall departments (unofficial departments such as Land Registry, Forestry Commission and Food Standards), local authorities and devolved governments. So, here are several interesting projects and experiments taking place in the UK:

✓ **Welfare distribution via blockchain technology**: the Department for Work and Pensions has teamed up with Barclays, RWE, GovCoin and the University of London in an experiment that will use blockchain technology to distribute welfare with a phone app. The trial was designed to see if payments could be sent and tracked using this technology.

✓ **DLT**: Credits, a Blockchain-based platform provider, and the UK government are collaborating on a framework that allows UK government agencies to experiment with Blockchain technology. (DLT stands for distributed ledger technology).

✓ **International payments based on Blockchain technology**: Santander bank has launched a trial of international payments based on

---

[36] UK Government Chief Scientific Adviser, (2016), "Distributed Ledger Technology: beyond block chain", Open Government Licence © Crown copyright, 87 p., available at https://goo.gl/asIz6L

Blockchain technology. Staff/staff in the pilot program involve an app that connects to Apple Pay. Users can use Touch ID to transfer payments between £10 and £10,000.

✓ **Using Blockchain technology to market gold:** the Royal Mint has teamed up with CME Group, a market operator, to use Blockchain technology to build a gold market in the hope of making London a more attractive city for gold sales. The technology is being adopted by both entities, as they see it as an efficient digital mechanism for gold trading.

### 7.3.2. The Singapore experience

Singapore, like the UK, has initiated attempts to experiment with Blockchain technology with the aim of increasing its financial attractiveness and making work just as easy.

The Monetary Authority of Singapore[37] (MAS) published its guidelines for a "regulatory sandbox" in 2016 to encourage and enable experimentation with solutions using technology in innovative ways for financial products or services. The term sandbox means a secure environment where developers can create software.

Singapore is taking steps to explore Blockchain technology, and it's paying off. One Singaporean bank, OCBC, has been using the technology for cross-border transfers. It has sent money to its subsidiaries, OCBC Malaysia and Bank of Singapore.

Singapore's central bank has also launched a pilot project, with eight foreign and local banks and the stock exchange. This proof-of-concept project aims to use Blockchain technology for its interbank payments. The pilot project also aims to examine cross-border currency transactions.

It's not just Blockchain companies that will be experimenting in Singapore. All the biggest players[38] have been involved, such as Bank of

---

[37] "MAS Issues "Regulatory Sandbox" Guidelines for FinTech Experiments", available at https://www.mas.gov.sg/news/media-releases/2016/mas-issues-regulatory-sandbox-guidelines-for-fintech-experiments;

[38] "MAS, R3 and Financial Institutions experimenting with Blockchain Technology", available at https://www.mas.gov.sg/news/media-releases/2016/mas-experimenting-with-blockchain-technology ;

America, Merrill Lynch, IBM, Credit Suisse, Bank of Tokyo-Mitsubishi UFJ Ltd, DBS Bank Ltd, JP Morgan, Hong Kong and Shanghai Banking Corp Ltd, OCBC Bank, United Overseas Bank, and the Singapore Stock Exchange.

Singapore already has a robust, modern digital identity system that could easily be connected to a blockchain.


### 7.3.3. The Dubai 2020 initiative

The Dubai government has set an ambitious plan to transfer all government documents and systems via Blockchain technology[39] . The dematerialization program is part of its initiative to become a world leader in this technology and boost efficiency across all sectors.

The new system will allow users to update and verify their credentials via the blockchain. They will only need to log in with their credentials once to gain access to government and private entities, such as insurance companies and banks. They also plan to share their technology with other countries to simplify border crossings.

Instead of passports, travelers could use pre-authenticated digital wallets, as well as pre-approved identification. The Dubai government has estimated that its initiative to adopt Blockchain technology has the potential to save 25.1 million hours of productivity. This boost in efficiency will also reduce carbon emissions.

In this same concept, the Dubai General Blockchain Council or Global Blockchain Council (GBC)[40] has announced seven new public-private collaborations, combining the skills and resources of startups, local businesses and government departments. They will apply Blockchain technology to the following:

✓ **Healthcare**: Estonian software company Guardtime will collaborate with one of Dubai's largest telecoms operators, Du, to provide the

---

[39] Smart Dubai Office, (2017), "Dubai -the first city on the Blockchain", case study© Smart Dubai, 19 p.

[40] "Dubai's Global Blockchain Council Announces Seven Pilot Projects", available at https://www.the-blockchain.com/2016/06/12/dubais-global-blockchain-council-announces-seven-pilot-projects-welcomes-new-members/;

technological expertise needed to digitize healthcare records and transfer them to blockchain.

✓ **Securities transfers:** securities transfers will be digitized and recorded on a Blockchain. A Singaporean Blockchain startup known as "Dxmarkets" has developed a proof of concept.

✓ **Business registration:** the GBC is experimenting with the use of Blockchain technology for business registration. This is different from Ethereum's Decentralized Autonomous Organization (DAO) but could streamline identity verification through the FlexiDesk program. It is currently in the demonstration phase, with several entities working on a proof of concept.

✓ **Tourism:** 'Dubai Points' is a pilot program that has been launched in collaboration with Loyyal, using Blockchain technology to help the tourism industry. It aims to encourage travel by awarding points to travelers who visit certain locations. It will use smart contracts to facilitate rewards. These points work a lot like a crypto token and can be redeemed.

✓ **Shipping:** IBM is working with the GBC with the aim of using Blockchain technology to improve shipping and logistics. The program aims to help regional players collaborate on how they exchange goods. Smart contracts will be used as solutions to compliance and settlement issues.

✓ **Diamond trade:** a pilot project will use Blockchain technology to authenticate and transfer diamonds. The Dubai Multi Commodities Center will digitize Kimberly Certificates, a certification system for the international trade in rough diamonds created by the UN to restrict the trade in conflict diamonds.

On the other hand, IT vendors such as HPE, IBM, Microsoft and SAP are now offering a new type of Blockchain-based service. This is 'Blockchain as a service (BaaS)', a solution that enables blockchain-based applications to be deployed without the need to invest in new infrastructure, or resort to Blockchain experts, who are in short supply on the job market. Table 2 shows a comparison of existing Blockchain-based service offerings.

| Publisher / solution | Launch year | Supported platforms | Pricing model | References |
|---|---|---|---|---|
| **HPE** | Early 2018 | R3 Corda | | |
| **IBM / IBM Blockchain** | February 2016 | Hyperledger Fabric version 1.0 | 752 euros per month | Bank of Tokyo, Mitsubishi UFJ, Northern Trust |
| **Microsoft / Blockchain on Azure** | November 2015 | Ethereum, Hyperledger Fabric, R3 Corda, Chain Core... | Linked to storage, compute and cloud services consumption | |
| **SAP / Leonardo Blockchain** | May 2017 | Leonardo program | | |

**Table 2Comparison of Blockchain as a Service offerings[41]**

It was Microsoft that first began offering a dedicated Blockchain service via its Azure cloud. This offering is tailored to a wide range of distributed ledger platforms such as Ethereum, Hyperledger Fabric....

For its part, IBM presents itself as the vendor with the most comprehensive BaaS solution. Its offering is based on an open source project from the Linux Foundation called Hyperledger Fabric[42] . With this choice, IBM is focusing on business-to-business networks capable of up to a thousand transactions per second.

Other players such as HPE (Hewlett-Packard Enterprise) and SAP (Systems Applications and Products in Data Processing) have also initiated projects in this field. The aim is to set up a decentralized registry to simplify the management of complex transactions between various stakeholders.

However, the idea of centralizing a Blockchain within a single service provider undermines the main idea behind Blockchain technology, which

---

[41] "Blockchain as a Service: which solution should you choose to get started?", available at https://www.journaldunet.com/solutions/cloud-computing/1206955-blockchain-as-a-service-quelle-solution-choisir/;

[42] "Hyperledger Fabric," available at https://www.hyperledger.org/projects/fabric ;

is the development of a decentralized platform without a trusted third party.

## 8. Opportunities and challenges of Blockchain technology

As discussed in previous sections, numerous use cases using Blockchain technology have been proposed, built and tested. The best way to discuss the effectiveness of an application or technology is to understand its opportunities and challenges.

### 8.1. Blockchain technology opportunities

When data is permanent and reliable in a digital format, you can carry out online transactions in a way that, in the past, was only possible offline. Everything that has previously remained analog, including property rights and identity, can now be created and maintained online. Slow commercial and banking processes, such as fund transfers and settlements, can now be carried out almost instantaneously. The implications for secure digital records are enormous for the global economy.

The first applications created were designed to piggyback on the secure transfer of digital value that Blockchains enable to be exchanged with their native tokens. These included things like the transfer of money and assets. But the possibilities of Blockchain networks go far beyond the movement of value.

Blockchain applications are built around the idea that the network is the arbiter. In this type of system, computer code becomes law, and rules are enforced as written and interpreted by the network. Computers do not have the same prejudices and social behaviors as humans. So the network cannot interpret intent (at least not yet). Insurance contracts arbitrated on a Blockchain have been extensively studied as a use case built around this idea.

Another interesting thing that blockchains enable is impeccable record-keeping. They can be used to create a clear chronology of who did what and when. Many industries and regulators spend countless hours trying to assess this problem. Record keeping using Blockchain technology will relieve some of the burdens that are created when we try to interpret the past.

What's more, the decentralized nature of this technology, coupled with its security and transparency, could potentially prove capable of replacing institutions, and promises applications that go beyond the monetary realm. Three categories can be distinguished[43] :

- Applications for transferring assets (money, securities, votes, shares, bonds, etc.)
- Blockchain applications as a registry to ensure better traceability of products and assets.
- Smart contracts, which are autonomous programs that automatically execute the terms and conditions of a contract, without the need for human intervention, once started.

The opportunities offered by Blockchain technology are mainly linked to the technological aspects presented in the previous sections. By eliminating intermediaries, the cost of money transfers can be reduced, for example, bank commissions cease to exist. Transfers can also be made more quickly, as crypto currencies are moved directly from one wallet address to another without intermediate steps.

Smart contracts offer a high degree of automation. Transparency is also guaranteed, as Blockchain technology is accessible by all network members. What's more, since anyone could potentially write to the register, Blockchain technology could become the repository for an enormous amount of information that could be used for data analysis in different sectors (insurance, finance, medicine, education, etc.). The underlying cryptographic mechanism guarantees that data is not altered and that transactions cannot be repudiated. Finally, the replication of the blockchain on each network node guarantees that the blockchain will survive unexpected events.

In addition, other opportunities may be linked to the ability of the market or public administration to adopt the technology or not. At present, interaction with Blockchain technology requires certain technical skills, such as mastery of the block concept.... But several efforts are being made

---

[43] "What is blockchain?", https://blockchainfrance.net/decouvrir-la-blockchain/c-est-quoi-la-blockchain/; accessed November 04, 2017

to reduce or mask the complexity of this technology. Companies or startups providing applications and services based on Blockchain technology could have a competitive advantage. This advantage would become even more important as IoT becomes more widespread.

Another opportunity is linked to the possibility of tackling new markets or sectors and creating new types of services such as DAOs. In this way, Blockchain technology could be successfully used to support the sharing economy.

On the other hand, if a large number of players write data onto the Blockchain, countless new applications could emerge. For example, a person's medical history could be easily retrieved by doctors in an emergency; indeed, Blockchain technology could become a repository of medical data that could be used by scientific researchers;

In addition, Blockchain-based supply chains could be more efficient, as data could be shared almost instantaneously between heterogeneous players; so in the case of an insurance scenario, for example, data could be used for fraud prevention, policy personalization and so on. Nevertheless, the type and impact of these applications would depend on the quantity and quality of the information recorded.

However, these promises are not without their challenges, be they economic, legal, governance or ecological...

## 8.2. The challenges of implementing Blockchain technology

The biggest challenges in implementing blockchain technology relate to scalability, energy consumption and performance. Indeed, at present, the number of transactions that can be processed per second is extremely low compared to traditional systems, and this is mainly due to the computing power required to validate new blocks. In this respect, it should be pointed out that some platforms based on Blockchain technology change the block validation process, reducing the complexity of the mathematical problem to be solved and limiting the possibility of exploration to only a subset of approved nodes.

Apart from the temporal context, space is also an issue, since data is replicated on every network node. As a result, the amount of energy consumed by network nodes is enormous, and the cost of the hardware required to validate new blocks is extremely high. For example, the Bitcoin blockchain requires over 170 GB of storage on each network node[44] , which is estimated to cost around $6 per transaction[45] .

On the other hand, the fact that the information encoded in the Blockchain is immutable and accessible to all, is another weakness that could harm users' privacy. To give an example, anyone can check how much money a person holds, by analyzing their incoming transactions. If other types of information are stored on the blockchain, such as medical records, this issue would become even more relevant. To address privacy concerns, some solutions have been proposed to anonymize payments or transactions.

The immutability and self-execution of code could be another challenge for Blockchain technology. Indeed, smart contracts could become easy to intercept and modify by hackers[46] . Indeed, hackers could exploit bugs in smart contracts to steal money, as was the case with the most famous attack of this kind on the Ethereum network where around $50 million was stolen in June 2016.

What's more, even if we assume that smart contracts are bug-free, some applications will still need external oracles to inject information into the Blockchain. Indeed, "the Oracle, is a service responsible for manually entering external data into the Blockchain"[47] . So, in this case, the weakest point would be the oracle. To remedy the consequences of injecting erroneous information into a Blockchain-based system, we could rely on more than one oracle, each obtaining information from different sources. Each oracle must prove its method of collecting information, either by

---

[44] "Cryptocurrency statistics", available at https://bitinfocharts.com/ ;

[45] "Let's quit the blockchain magic talk," available at https://www.zdnet.com/article/lets-quit-the-blockchain-magic-talk/ ;

[46] "What is a DAO?", available at https://blockchainfrance.net/2016/05/12/qu-est-ce-qu-une-dao/ ; And "The DAO: a hacker steals $50 million, the counter-attack prepares", available at https://www.nextinpact.com/news/100336-the-dao-pirate-derobe-50-millions-dollars-contre-attaque-se-prepare.htm ;

[47] "Oracles, the link between blockchain and the world," available at https://www.ethereum-france.com/les-oracles-lien-entre-la-blockchain-et-le-monde/ ;

using a provable-honest service or by using a consensus-based oracle in the process of collecting and sending information[48] .

In addition, other challenges affecting the usability of blockchains can be identified. Firstly, the inability to receive assistance in the event of lost credentials. This weakness could be partially eliminated by relying on trusted services.

Another challenge is that development tools are still at an early stage, and standards for the development of applications based on Blockchain technology have yet to be defined. Finally, it is interesting to note that, in some cases, Blockchain would not be the most appropriate technology, as existing, well-mastered alternatives would deliver comparable results[49] .

Other challenges are linked to various external causes. Indeed, there is always a risk that some public or private players will be wary of this technology, considering it insecure or unreliable, due to bugs for example... Other players might think it's too complicated, and that the rate of adoption on a global scale could be low.

Particular attention should be paid to questions relating to legal regulations, which could threaten the adoption of Blockchain technology, such as those concerning the legality of smart contracts or the dissemination of personal data. For example, the regulatory framework is not yet ready to protect the system's transactions.

What's more, applications based on Blockchain technology represent a medium- to long-term investment, and they can't be integrated into all existing processes.

Finally, we could be faced with a refusal to adopt this technology on the part of individuals, as they may still consider personal interaction to be important. So, a communication policy should accompany any change based on this technology.

---

[48] "Oracles, the link between blockchain and the world," available at https://www.ethereum-france.com/les-oracles-lien-entre-la-blockchain-et-le-monde/ ;

[49] "Do You Need a Blockchain?", available at https://spectrum.ieee.org/computing/networks/do-you-need-a-blockchain ;

## Conclusion

In this chapter, an overview of Blockchain technology was presented, as well as potential applications and use cases for this technology. The opportunities and challenges of this technology were also discussed.

Blockchain technology is receiving increasing attention from research, industry and governments, and is regarded as a revolutionary technology. However, the key problem reported by several public and private players is the lack of an objective assessment of whether or not to invest in this technology. Indeed, the risk lies in the fact that a stakeholder decides to adopt Blockchain technology because it is the trend, without asking whether it is mature enough to be adopted in everyday activities and by a wider audience. To this end, in the following sections, we'll try to frame this technology in the Tunisian context.

# Chapitre II.  Blockchain and online services

## Introduction

In this chapter, we try to present, on the one hand, the coupling between e-government and Blockchain technology in a global way (1), and on the other hand, between the latter and Tunisian public services (2). In a second section, we discuss the guidelines under which some Tunisian public authorities have become involved (3). This is followed by a review of the Tunisian Post's experience in developing digital administration using Blockchain technology (4). Finally, all of the topics discussed will help to frame the technological choices needed to develop the possibilities and potential for integrating this technology into Tunisian public services (5).

## 1. Blockchain technology and e-administration

In this section, we will focus on the challenges of e-government (1.1) and its relationship with public administration (1.2).

### 1.1. The challenges of e-administration

Electronic government[50] , or e-government in English, is the emergence of a new channel of communication, driven by ICT, with improved services and greater transparency. The concept began to emerge in the 80s[51] , with extensive reforms carried out by governments in Western and European countries, and by international organizations such as the OECD.

---

[50] United Nations Department of Economic and Social Affairs, (2014), "United Nations E-Government Survey 2014", Copyright © United Nations, page 2

[51] Gauche K. and Taddei R., (2011), "Enjeux et services de l'administration électronique locale. Etude de cas", Nouveaux usages de l'internet dans les collectivités territoriales, IAE NICE, France. 19 p.

The aim of e-administration is to respond to the spread of an information society, which is developing new needs on the part of citizens, such as accessibility, speed and simplification of procedures.

In France, for example, e-government has become a central plank of public policy, with an increasing number of modernization programs, such as the "service-public.fr" portal, aimed at improving the quality and efficiency of administrative procedures and the services provided to citizens.

For its part, Canada is one of the most advanced countries in terms of e-administration, with a single portal containing more than one hundred and thirty-five administrative services accessible online, corresponding to the equivalent of eight million paper forms.

There are four stages in the dematerialization of information or transactions[52] . The first stage is characterized by putting information online for consultation purposes. The second stage involves downloading a form for printing, without the option of filling it in online. Then, the third stage enables the dematerialized sending of a form dealing with an online request or declaration. And finally, the fourth stage is characterized by complete dematerialization, via the acquisition of an account and personal monitoring space.

For the user, e-government should make administrative services more responsive, reduce service costs and improve quality. This can be translated into four criteria[53] which are simplicity, access to global services by encouraging collaboration between the various administrative components that are currently compartmentalized and bureaucratized, multi-channel access via various tools and the creation of new services based on multiple data such as geolocalized data.

In this way, e-administration represents a new relationship between the State, its administrative structures and citizens. However, the integration of the citizen as a contributor is essential to propagate the

---

[52] United Nations Department of Economic and Social Affairs, (2014), "United Nations E-Government Survey 2014", Copyright © United Nations, page 195

[53] Gauche K. and Taddei R., (2011), "Enjeux et services de l'administration électronique locale. Etude de cas", Nouveaux usages de l'internet dans les collectivités territoriales, IAE NICE, France. 19 p.

notion of open collaborative innovation and accelerate the desired modernization. What's more, the open data approach adopted by public administration is an important tool for fostering the notion of proximity with citizens, by facilitating their knowledge of and access to their own data or other data that encompasses their environment and daily life.

As a result, e-administration is a factor in the evolution of administrative culture towards the renewal of the relationship between the State and its citizens, by introducing new relational parameters of proximity and responsiveness.

But the digitization of administrative services presents a number of challenges, such as the existence of numerous operating rules, the diversification in quantity and quality of the data manipulated by these services, and the associated security. For this to happen, the State needs to reach a consensus among the various players in the administrative ecosystem to overcome this technical complexity, in addition to the ever-present political stakes.

The progress of e-administration also depends on the State's ability to reduce the gap in terms of the digital divide between the various regions of which it is made up. Thus, the processing of public data requires collective reflection and legal treatment to ensure the sustainability of any technical solution adopted.

E-administration fosters e-democracy through exchange and reflection on the frequent issues[54] at the heart of e-administration, such as exchange security, open data, personal data protection and the State's digital sovereignty.

The deployment of[55]online services, and consequently the construction of an online administration, can be carried out in several stages, ranging from putting information online, to putting forms online, and arriving at the generalization of a tele-services offer. This deployment

---

[54] Brown D., (2005), "E-government and public administration", Revue Internationale des Sciences Administratives (Vol. 71), @ I.I.S.A., pp. 251-266, available at https://www.cairn.info/revue-internationale-des-sciences-administratives-2005-2-page-251.htm

[55] Oberdorff H., (2006), "L'administration électronique ou l'e-administration", In: Recherches et Prévisions, n°86, La nouvelle administration. L'information numérique au service du citoyen, p. 9-18; available at http://www.persee.fr/doc/caf_1149-1590_2006_num_86_1_2247

promotes the construction of a common base of shared applications, accelerating digital transformation, especially at territorial level, and guaranteeing shared governance between the State and its local authorities, promoting a global approach to data management in the service of policies of general interest at local, regional and national level.

## 1.2. Relationship between e-government and Blockchain technology

E-government[56] is an area of significant change through innovation and continuous development. It varies according to the institutional regime applied, the nature of government activity, the objectives governments seek to achieve, and the specific tools that are built and deployed. The coupling between e-government and Blockchain technology would enable public administration to be repositioned around several principles such as:

- Citizen-centric service: Blockchain technology would enhance a fundamental principle of e-government, namely citizen-centric service delivery. Indeed, through this principle, the creation of public services should be based on the overall objective of satisfying citizens' needs. This would foster the emergence of the notion of self-service, where citizens could carry out a large number of administrative tasks via the various channels of communication that used to exist between the administration and its public.

- Information as a public resource: Blockchain technology would reinforce the crucial value of information collected or produced within e-government. Indeed, information is seen as a key resource for the administration to ensure the sustainability of its actions and interventions towards the citizen. We are therefore faced with a bank of administrative information that encompasses several processes, such as collection, production, use, storage, extraction, processing, dissemination, protection and deletion, and which requires clear legislation and dedicated institutions.

---

[56] Henman P., (2010), "Governing Electronically: E-Government and the Reconfiguration of Public Administration, Policy and Power", First published by Palgrave Macmillan, 280 p.

- Relational renewal: Blockchain technology would promote networking. Indeed, this method of working is based on collaboration and information sharing between managers and civil servants in administrative departments, in addition to constituents. This would enable new relational management mechanisms that go beyond the bureaucratic vertical division of labor. In addition, the State and its administrative structures need to develop partnerships with the national and multinational private sector, which has experience in innovation and wealth creation through this new technology.

In conclusion, e-government coupled with Blockchain technology would change the accountability and management models of traditional administration by influencing its role in promoting economic development and social cohesion, its relationship with its constituents, its internal organization and its relationship with the external national and international environment.

## 2. Blockchain technology and online services in Tunisia

The modernization and consolidation of governance in public structures currently represents a major challenge for our country, given the acceleration of political and social change, the globalization of trade and the development of new information technologies.

As a result, Tunisia began to realize that Blockchain technology could be used for a multitude of tasks, including providing innovative ways of interacting with citizens and developing digital public services.

Renovating the public service and gearing it towards performance and quality delivery to users requires profound changes that can be achieved by coupling with Blockchain technology, which is what is discussed in this section.

### 2.1. E-administration as a major challenge for public services

In the information society, the needs, uses and quality standards of users have evolved. Faced with these changes, and since the private sector has developed considerably to respond to these new trends, public administrations and local authorities are obliged to do as much as they can to keep up with the movement and satisfy citizens. To achieve this, electronic administration or e-administration must be at the heart of their concerns.

The challenge is to exploit technological opportunities to ensure a close relationship with customers, enhance the effectiveness of the actions proposed by the various state structures and improve the quality of public service.

In a context characterized by major political, social and economic change, Tunisia is aware of the challenges that can be brought by e-government. To this end, in 2018 it defined an e-gov strategy named 'SmartGOV 2020'[57] which revolve around a major priority of becoming a global digital reference while making ICT an important lever for increasing socio-economic development. The aim was to improve the development of e-government as part of an overall strategy to reform and modernize the organizational, legal, technological and human levers of government.

In an interview[58] in December 2017, Abdelkader Labaoui, Director General of the National School of Finance, said that e-government will generate financial revenues of up to 62 billion dinars over an estimated period of 10 to 15 years. According to Labaoui, this amount is equivalent to 50% of Tunisia's GDP.

E-administration certainly makes it possible to create new perspectives, improve the transparency of actions and better gather the needs and personalized profiles of users; moreover, it enables the provision of personalized services to better segment the offer to users, as illustrated in figure 3, and facilitate their access to public services. It is therefore essential to develop a relevant offer to provide any user with a modern, legible, quality public service at the lowest possible cost. This can be achieved by adopting an approach that involves developing a more

---

[57] "Overview of e-strategy", available at http://fr.tunisie.gov.tn/; accessed August 13, 2018

[58] "E-government will mobilize 60 billion dinars," available at https://www.leconomistemaghrebin.com/2017/12/28/administration-electronique/; accessed August 13, 2018

citizen-centric approach and including new technologies such as Blockchain.

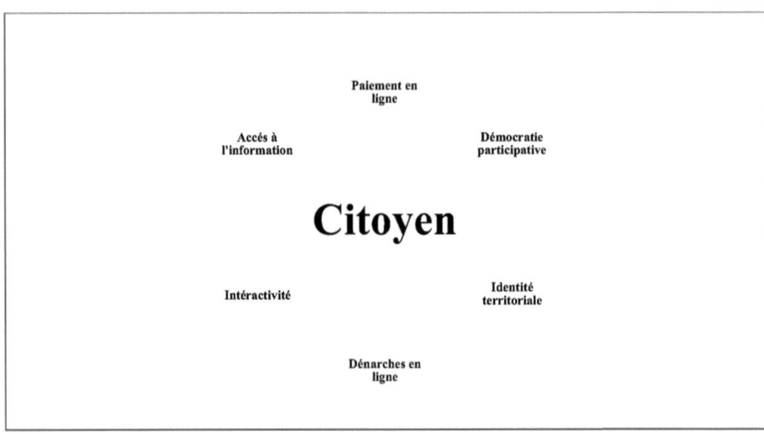


**Figure 3:Personalizing citizen services**

Admittedly, the development and generalization of e-administration through Blockchain technology will bring efficiency and fluidity to the practices and operation of public structures. But this development is hampered by resistance to change due to the cultural and sociological context in addition to the heterogeneity of the system, which makes it difficult to appropriate the new uses of e-administration within public services themselves.

Indeed, to achieve dematerialization, each department or public structure must review its information system and working methods. To this end, administrative staff will have to change their ways of thinking, learning, working and cooperating to keep pace with this movement and evolution.

### 2.2.Blockchain technology at the service of public action

A large number of professions and services require registers that are unalterable, consensual and public. Before the advent of digital technology, the solution was to have a centralized authority certify inalterability and provide a guarantee equivalent to consensus. This role was therefore generally entrusted either to governments or to private operators, most often subject to public control.

Blockchain technology offers the possibility of decentralized construction of an attack-resistant register, recording data in a definitive and shared way, with a consensus of users or operators on the content of this register.

In the case of public action, the decentralization of databases represents a paradigm shift for the usual organization of public data hitherto stored by one or more closely supervised public operators.

Indeed, the main features of blockchain technology are of interest to public action. A blockchain is essentially a method of implementing a distributed ledger that is protected against changes to the data stored, including by those who implement it. Technically, the fundamental properties of blockchains can be of interest for public action: thanks to this technology, citizens can be offered access to a high level of transparency, while at the same time being able to actively participate in the system implemented.

Thus, Blockchain technology appears to be a means of refounding certain modes of operation by conferring on citizens or a subset of citizens a power of verification, thereby reinforcing the trust that the data concerned can generate, which is today reserved for the public authorities.

This evolution will not only improve the fluidity of exchanges, but also facilitate the control of operations, leading to the efficient operation of public services.

### 2.3. Blockchain ecosystem

Each Blockchain is built for its own purpose, and for each purpose there are a number of added elements illustrating specific rules of use. Seen in this light, a[59] Blockchain ecosystem represents independent parts that interconnect. Thus, a strong Blockchain ecosystem will have

---

[59] "What Do We Mean When We Talk About the "Blockchain Ecosystem"?", available at https://bitcoinmagazine.com/articles/op-ed-what-do-we-mean-when-we-talk-about-blockchain-ecosystem;

components that work harmoniously together to provide a quality purpose with seamless interconnections.

In general, a Blockchain ecosystem comprises four main parts, illustrated in Figure 4: the Blockchain layer, the technology layer, the Blockchain-service layer and the user layer.

The first layer, the blockchain layer, contains the various blockchain platforms, through which the users and transactions of a given service are recorded. Each platform may have specific technical features that favor particular types of application.

The second layer is the technology layer. It encompasses companies acting as a technical interface between a Blockchain and the services they offer. They enable the information contained in a Blockchain to be processed for use by third-party services.

The third layer is the Blockchain-service layer, whose role is to define the set of applications used directly by the end-user. The technical operation of these applications is always transparent.

Finally, the last layer of the Blockchain ecosystem is the layer of users, who are those who benefit from the services provided by Blockchain technology, such as state structures, private players, associations and individuals.

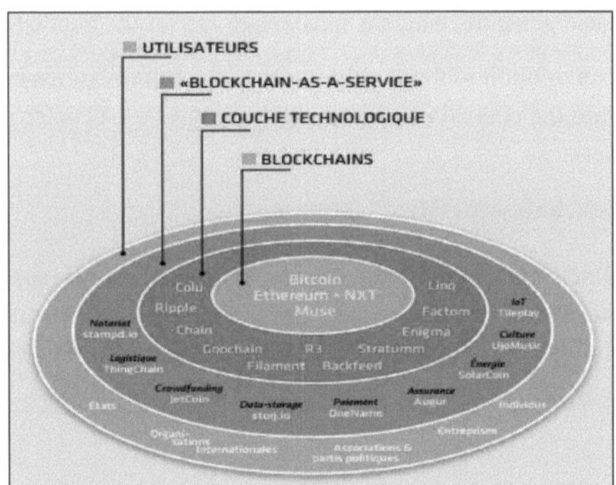

**Figure 4Blockchain ecosystem[60]**

## 3. Policy instruments and guidelines

Tunisia has not lacked interest in integrating new technologies into its economic, social, administrative and governmental fabric. This explains the interest in Blockchain technology, which has become an international technological trend.

There are several indications of the willingness of Tunisian government structures to take advantage of this technology. The first appears from the fact that Tunisia organized a conference in May 2018 entitled the "Africa Blockchain Summit" under the leadership of the Central Bank of Tunisia, in collaboration with the representative organization of the Paris financial marketplace "Paris EUROPLACE"[61] and the technical support of the Talan Group[62] . A conference on the theme of "Blockchain technology and the prospects it offers for banking and finance"[63] , provided a forum for dialogue and exchange of experience on the advances and uses of Blockchain technology.

The technology summit was attended by all African central bank governors, multinational institutions, African financial players, academics and researchers.

Indeed, many players such as companies, governments, etc. are considering the use of this technology for other use cases other than digital currency. This initiative offers an opportunity to contribute to the reflections and work being carried out worldwide on Blockchain, and to propose, a common regional vision of its use.

---

[60] Guillaume B., (2016), "Understanding Blockchain - anticipating Blockchain's disruptive potential on organizations", uchange.co, page 28

[61] "Presentation of the Paris EUROPLACE organization", available at www.paris-europlace.com/fr ; accessed June 10, 2018

[62] "Talan Group presentation", available at https://www.talan.com/; accessed June 10, 2018

[63] "BCT press release for the kick-off regarding the organization of the "Africa Blockchain Summit"", available at https://www.bct.gov.tn/bct/siteprod/actualites.jsp?id=413 ;

At the opening of the "Africa Blockchain Summit", the Governor of the Central Bank of Tunisia confirmed that "The Tunisian monetary and financial community is convinced that the new Blockchain technology is a source of growth for Tunisia as well as for all of Africa." [64]

On the other hand, the BCT governor pointed out that African and Arab countries face common challenges such as low financial inclusion, a vast informal sector that accounts for 85% of all jobs in Africa and 58% in Tunisia, according to the latest study by the International Labor Organization (ILO), high unemployment, particularly among women, young people and higher education graduates, insufficient loan guarantees, a lack of infrastructure and significant demographic expansion.

As a result, Blockchain technology could be a tool enabling countries that adopt it to accelerate their economic emancipation and perpetuate their autonomy, which would go beyond the monetary aspect.

However, according to the BCT governor, there are several challenges to the development of this technology. The first is the lack of connectivity in Africa. The second obstacle is that any development of a Blockchain application requires preparation for a transition phase, as the solutions offered require a major change or even a complete break with existing systems.

Finally, the third challenge depends on the involvement of the public authorities in legislating this technology so that it can be used on an international scale, while ensuring the stability of financial rules, the protection of users and their personal data, and the fight against money laundering.

On the other hand, there is a second indication that Blockchain technology is at the heart of the concerns of Tunisian public structures. This is the "Code4Chainge, Africa's largest Blockchain hackathon"[65] ,

---

[64]     "Live     stream     Africa     Blockchain     Summit",     available athttps://www.bct.gov.tn/bct/siteprod/actualites.jsp?id=485&la=AN;

[65] "Talan organizes Code4Chainge, Africa's largest blockchain hackathon", available at https://thd.tn/talan-organise-le-code4chainge-le-plus-grand-hackathon-blockchain-africain/;

organized in conjunction with the African Blockchain Summit. Code4Chainge is a competition in which teams of researchers and developers compete for 36 hours non-stop around a Blockchain challenge.

During the competition, the Banque Centrale de Tunisie team was one of two winners[66]. It presented an innovative banking project that controls the inflow and outflow of cash in African countries via a platform linking banking institutions to the Central Bank.

In 2020, the Central Bank of Tunisia unveiled its first "regulatory sandbox", a mechanism that enables fintechs to launch experimental financial services in a test environment to support the experimentation of innovative solutions on a small scale and with real customers. The test phase of the first cohort of this regulatory sandbox[67] has been launched with four selected fintechs offering solutions related to remote customer identification, cross-border clearing, cryptocurrency and digital central banking based on cutting-edge technologies such as artificial intelligence and blockchain.

Just over three years later, the BCT has decided to replicate this experience, but this time with more "traditional" financial institutions, by launching the "sandbox express"[68] for local banks. Through this initiative, banks will have the opportunity to launch and test their digital solutions to ensure compliance with the various regulations. For the BCT, this framework offers an opportunity to verify the robustness of these solutions before they are made available to users.

Finally, other attempts to use Blockchain technology by public authorities in Tunisia have already been made, as in the case of the Tunisian Post Office, or are in the process of being deployed, as in the case of the Ministry of Higher Education. The following sections detail these use cases.

---

[66] "Tunisia: Two teams win the final of the "Code4Chainge" Blockchain Hackathon", available at https://www.ilboursa.com/marches/tunisie-deux-equipes-remportent-la-finale-du-hackathon-blockchain--code4chainge-_14193%22%3E ;

[67] Launch of a "Regulatory Sandbox" with the BCT available on https://fintech.bct.gov.tn/fr/acualite/lancement-d%E2%80%99une-%C2%AB-sandbox-r%C3%A9glementaire-%C2%BB-aupr%C3%A8s-de-la-bct

[68] SandBox express available at https://fintech.bct.gov.tn/fr/sandbox-express

## 4. Study of the experience of the Tunisian Post Office

In Tunisia, La Poste, the leader in financial and social inclusion, is still on a trend to promote various digital services in Tunisia. In October 2015, La Poste had proceeded to launch the new Blockchain experience in partnership with Tunisian Fintech startups.

Tunisia will thus become the first country in the world to operate its national currency via the Blockchain. The Blockchain in question is an advanced version of Bitcoin's developed by Monetas, a Swiss start-up[69] .

With a population of around 11 million, Tunisia has over 3 million unbanked people[70] . However, 658,725[71] of them use the e-Dinar, a digital version of the national currency via La Poste tunisienne, which is the only non-producing entity authorized to collect savings. La Poste Tunisienne has been providing accounting services since 1918, while the national savings bank was created as early as 1956.

The platform, developed on behalf of Poste Tunisienne and in partnership with startups[72] Monetas and DigitUs, uses Blockchain to enable a variety of transactions worldwide. Indeed, the aim is to be able to replace the self-created e-Dinar electronic currency with a Blockchain version enabling Poste Tunisienne customers to issue instant money transfers and in-store and online purchases via QR codes[73] . This will also enable them to pay their bills, and even manage their government identity documents. Transaction costs will be negligible, and issuing and distribution operations will be controlled by the Post Office.

---

[69] "Presentation of the startup Monetas," available at https://monetas.net/ ; accessed June 03, 2018

[70] "Tunisia to Transfer its National Currency onto Blockchain," available at http://forklog.net/tunisia-to-transfer-its-national-currency-onto-blockchain/ ; accessed June 03, 2018

[71] La Poste Tunisienne, (2016), "Annuaire Statistique 2016", @Poste_Tn, 39 p.

[72] "Communiqué de la startup Monetas concernant le partenariat avec la poste Tunisienne et la startup DigitUs", available at https://static.nzz.ch/files/7/5/3/monetas+tunesien_1.18629753.jpg;

[73] "Tunisia's Post Office Trials Crypto-Powered Payments App", available at https://www.coindesk.com/tunisias-post-office-trials-crypto-powered-payments-app/;

The idea of integrating Blockchain technology stems from the fact that the Tunisian Post Office is a very important and trusted institution, at the heart of Tunisia's financial inclusion efforts. As such, it is always in the midst of a transformation to modernize its services with innovative technologies and power the digital economy.

The platform deployed, whose operating architecture is illustrated in figure 5, is called "DigiCash" and consists of a mobile application enabling the creation of a virtual wallet that can be topped up with an e-Dinar account.

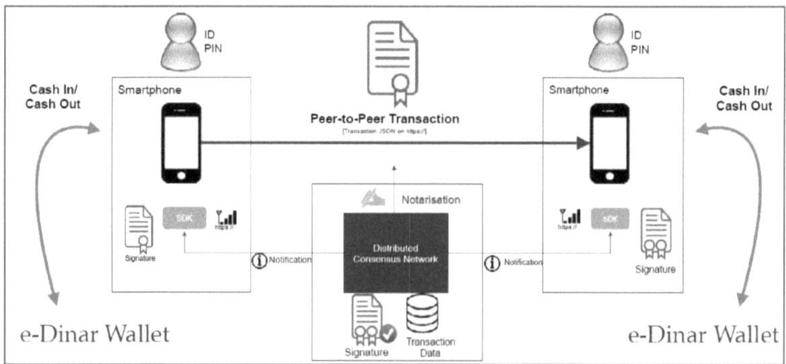


**Figure 5:Blockchain-inspired DigiCach transaction protocol[74]**

However, this kind of change in financial practices faces several regulatory challenges. Firstly, the Tunisian Post is an authorized financial institution, but not a bank. Secondly, the Tunisian currency TND is not a convertible currency, which limits the benefits of such an application for international financial exchanges. Finally, there are no legal rules officially authorizing the introduction and use of crypto-currencies in Tunisia.

For all that, we can get around some of these challenges if we consider, for example, that Blockchain is a technology and not necessarily a crypto-currency that might need to be regulated.

---

[74] Chakchouk M., (2017), "Blockchain in Tunisia: From Experimentations to a Challenging Commercial Launch", ITU Workshop on "Security Aspects of Blockchain" Geneva, Switzerland, 9 p.

Finally, we can say that Blockchain technology is an opportunity to foster digital financial inclusion through a profound and dynamic change in the Tunisian financial sector. However, its implementation must take into account the sovereignty requirements adopted by the Tunisian state, as well as the risks of IT security.

## 5. Potential for integrating Blockchain technology into Tunisian public services

The use of Blockchain technology in e-administration could lead to a revolution in Tunisian public services, given its ability to consolidate data and transaction security, reduce costs, simplify procedures and increase the transparency and speed of services. At present, conditions are favorable in Tunisia for the adoption of Blockchain technology, especially as the skills are available. In this section, we will focus on a few Blockchain-based application prototypes currently in the pipeline, as well as proposing new potential use cases for the benefit of Tunisian citizens.

### 5.1. Prototype in progress

At a global level, most central banks have launched initiatives and experiments. For example, the Banque de France has tested a decentralized SEPA (Single Euro Payments Area) creditor identifier register. The same goes for the Central Banks of Japan, China and Russia, which are testing Blockchain-based payment and settlement-delivery solutions, etc. These various developments are being monitored by global financial authorities such as the World Bank, the IMF and the BIS by publishing analytical reports and opinions.

On the Tunisian side, the Talan Group, in response to a request from Tunisia's central bank, is developing a check verification platform[75] . This use case, which is still in the prototype phase, will enable merchants to check checks received from their customers before cashing them with their banks. The aim of this tool is to enable merchants to check the

---

[75] "Blockchain, multiple use cases without cryptocurrency", available at https://thd.tn/la-blockchain-des-cas-dusages-multiples-sans-cryptomonnaie/ ;

various states of a cheque, such as its authenticity, the status of its issuer (in good standing or not), whether it has been stolen or not, etc....

Such an application will enable economic growth within the country, as well as any collaboration with the outside world, by establishing, on the one hand, the reliability of exchanges, and on the other, trust between the various players involved in this exchange process.

On the other hand, another prototype application integrating the Blockchain solution is currently being finalized. This is a platform for establishing and equalizing university diplomas[76] on behalf of the Ministry of Higher Education.

The advantage of such a solution is that it enables citizens, and young students in particular, to continue their studies abroad by validating their diploma with a foreign institution in no more than a few minutes. This requires collaboration between a number of players, including universities, public and private institutions, and the relevant ministry.

The deployment of blockchain technology is currently being finalized. It would touch the heart of financial institutions' information, organization and value-creation systems.

Similar to these attempts to deploy Blockchain technology, and with the aim of improving e-administration opportunities, we can propose new Blockchain platforms that would touch other areas of activity, and in which the citizen will be its center of concern.

### 5.2. Proposals for new sectors

We want to target areas that will affect as many users of public services as possible. We have therefore chosen to focus on two areas: civil status records and patient data management.

The aim is to mobilize local and national players around shared projects that will enable them to keep pace with the major transformations undergone by the State and local authorities.

---

[76] "Blockchain app for diploma legalization to be launched in October," available at http://www.gnet.tn/actualites-nationales/une-application-blockchain-pour-la-legalisation-des-diplomes-sera-lancee-en-octobre/id-menu-958.html;

In this way, we could provide keys to understanding, through Blockchain technology, about possible future transformations of computer systems to make them suitable for exchanging various important information and documents, saving them and distributing them to multiple network users without going through a central control structure.

### 5.2.1. Civil status documents

In this section, we attempt to describe the Madania national civil registration system before identifying the prospects for improvement through Blockchain technology with reference to Estonia's experience in the field.

#### 5.2.1.1.  Context

Since 1998, civil status records have been entered electronically at commune and rural center level, and since 2000, all the information contained in these records has been transferred to a central site in Tunis[77] . The national civil registry system, Madania[78] , manages all civil registry documents (birth, marriage, divorce and death certificates, legal texts, etc.) and publishes them at all the sites concerned (municipalities, municipal districts and commercial sites). The system thus enables:

- Electronic exchange of civil status documents with national social security funds ;
- Editing of civil status documents regardless of place of registration (remote);
- Data transmission and exchange between the user sites concerned;
- The creation of a reliable centralized database ;
- Serve citizens with civil status extracts not only from any municipal district dependent on the municipality holding the register of deeds, but also from any other district in any other Tunisian municipality.

---

[77] AIMF, (2004), "Fonctionnement de l'état civil dans le monde francophone", @ AIMF, page 38

[78] "The Madania national civil status system", available at http://www.cni.tn/index.php/fr/pages-3/item/149-madania;

A more advanced version of this system has been developed under the name "Madania2", with the aim of enabling institutions that frequently request civil status documents, such as social security funds and educational institutions, to have these documents available online without having to request them from the interested parties.

In addition, since May 2016, a new system for civil status[79] , whose home interface is shown in figure 7, has been put online by the Ministry of Communication Technologies and Digital Economy. This site makes it possible to apply for a birth certificate online, consult its progress and collect it via the Post Office for Tunisians in Tunisia or residents abroad, without having to resort to queues in the municipalities

For delivery in Tunisia, the total cost of a single copy of a birth certificate ranges from 3,850 dt for delivery by registered mail, to 5,500 dt for delivery by rapid-poste. These amounts are quite high compared with the price of a single traditional birth certificate delivered by a Tunisian municipality, which is around 500 millimes. This may not encourage Tunisian citizens to use this service. In addition, delivery times are also variable, and the extract can only be obtained directly and after identity verification on receipt.

Furthermore, in the case of a request for a birth certificate in French, the operation will only be processed if the translation has already been carried out previously in the municipality of birth of the owner of the birth certificate.

---

[79] "The online civil status platform", available at https://www.etatcivil.gov.tn/Madania/web/indexfr;

**Figure 6:Vital records platform interface[80]**

Blockchain technology has potentially infinite applications. But, logically, this technology is only useful if it can correct difficulties and make things better. So, what still needs to be improved for the civil registry?

On the one hand, interoperability between services is poor or non-existent. Today, for example, you can get a birth certificate from any municipality, but if the network malfunctions, you have to request your birth certificate from the municipality where you were born. The same applies to any kind of modification to the birth certificate.

On the other hand, while civil status has been modernized, there are still things to be improved. Such is the case with our identity card, which only serves to prove our identity when it could do so much more.

The national identity card (CIN) is an official document enabling all Tunisian citizens to prove their identity and nationality. It is governed by law N°93-27 of March 22, 1993 relating to the national identity card[81] , modified and completed by law N°99-18 of March 01, 1999. The CIN card is essential for exercising various rights, such as taking part in elections, completing administrative formalities and so on. It may be

---

[80] "The online civil status platform", available at https://www.etatcivil.gov.tn/Madania/web/indexfr;

[81] "Law No. 93-27 of March 22, 1993 on the national identity card.

required in the event of an ordinary identity check by law enforcement agencies in Tunisia.

As part of the promotion of a national electronic citizen identification system aimed at creating a national database for identifying citizens via a single national identifier enabling access to different systems and services, a law[82] on the biometric CIN has been passed with the aim of replacing the current identity card with a biometric ID document.

In fact, the biometric identity card is equipped with an electronic chip that can contain a range of personal information that will be centralized and consulted by the Ministry of the Interior. With a storage capacity of over 80 gigabytes, the card is expected to replace all the cards currently in use, such as CNSS and CNAM healthcare cards, driving licenses, etc., making administrative work much easier and simplifying access to various administrative services for citizens.

The biometric CIN card is used in various developed countries such as France, Belgium, etc... however, the debate that accompanied this bill, concerning the nature of the data on the electric chips, the conditions of access and storage of the information and the bodies, inside and outside the State, that should control and censor the use of the data collected in compliance with the legislation in force, ended up causing the Ministry of the Interior to withdraw the text in January 2018.

### 5.2.1.2. Blockchain technology:

Through Blockchain technology, distributed and decentralized registers would enable interoperability. By building a network based on Blockchain, all civil status records can be stored in a secure, forgery-proof way that can be consulted by anyone.

This results in a number of improvements, such as lower costs, the potential elimination of fraud and error risks, and the sharing of data on a distributed, non-centralized register.

As a direct consequence, Blockchain technology would also improve the usefulness of our identity card, which could be used for other purposes

---

[82] Organic law no. 2024-22 of March 11, 2024, amending and supplementing law no. 93-27 of March 22, 1993 on the national identity card

beyond civil status. This already exists in Estonia, however, which can be considered a reference model for any country.

### 5.2.1.3. Estonia, a model for all

The Estonian example of e-administration is absolutely impressive. This is a Baltic country, formerly Soviet, independent since 1991[83] and with a standard of living no higher than that of the European Union. This country has built the most modern state administration in the world. It has taken advantage of the technological innovations of the 90s to make dematerialization the basis of Estonian administration, in order to make administrative procedures as efficient as possible.

Almost all Estonians now have an electronic identity card, containing a chip that enables its holder to access a multitude of services. Indeed, in Estonia 99% of public services are accessible online, and there were 1,789 administrative services available online in 2016[84] .

Thus, thanks to a unique identifier, comparable to our CIN card number in Tunisia, each citizen can access, for example, his or her health data such as reimbursements, medical analyses, prescriptions... and also grant access to said data to a third party. In fact, the ID card chip contains cryptographic keys, including a private key that is the sole responsibility and possession of its owner. The identity card gives each citizen access to over a hundred public databases, as illustrated in figure 8 below.

---

[83] Digital Exploration, (2015), "Estonie se reconstruire par le numérique", (cc) creative commons - Renaissance Numérique, page 6

[84] "How Estonia is decentralizing its administration with blockchain", available at https://www.journaldunet.com/economie/finance/1197222-comment-l-estonie-decentralise-son-administration-avec-la-blockchain/ ;

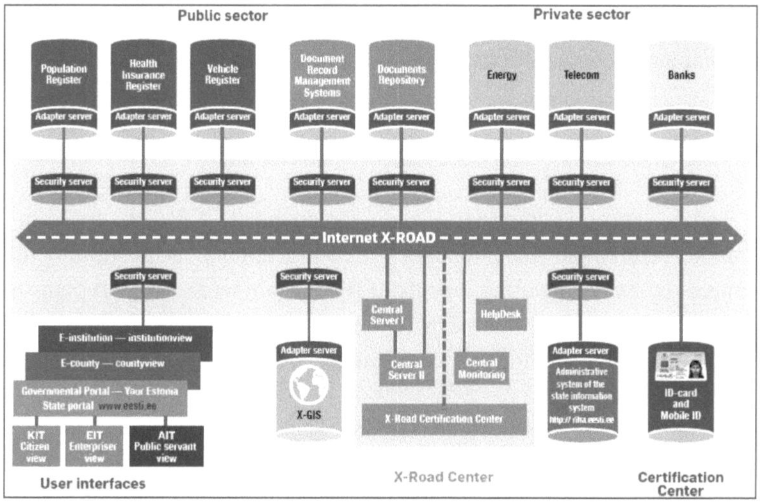


Figure 7Estonian information system[85]

In terms of security, and following a cyber-attack by Russian hackers in 2007, Estonia created the KSI "Keyless Signature Infrastructure" in 2008[86] , which is a true Blockchain where every piece of data is traceable, as it is entered into the register with a hash. This means that every consultation or attempt to consult the data is visible. The KSI Blockchain is an unforgeable distributed register, since there is no central server.

In Estonia, state institutions and departments are obliged to exchange information using the "X road" cryptographic protocol[87] . To ensure secure transfers, all outgoing data from X-Road is electronically signed and encrypted, and all incoming data is authenticated and recorded in the various information systems[88] . In this way, a user only needs to enter data once on a public service site, and all the other sites will automatically

---

[85] Anthes G., (2015), "Estonia: A model for e-government", Article in: Communications of the ACM (vol 58), N°6, pages 18-20

[86] Tallinn Entrepreneurship Office, (2017), "Tallinn in brief 2017", Tallinn - Estonia's economic center, page 14

[87] Digital Exploration, (2015), "Estonie se reconstruire par le numérique", (cc) creative commons - Renaissance Numérique, page16

[88] "Introduction of X-Road", available at https://www.ria.ee/en/introduction-of-xroad.html;

register the information. Once a set of information has been provided by an Estonian citizen to the state, many processes can be automated.

What's more, X-Road resides in data confidentiality, as everyone can see who is consulting their data and for what reason.

Today, several countries have implemented X-Road, including Finland, Azerbaijan and Namibia[89] .... And Estonia is continuing its progress by exploring other aspects of Blockchain technology. It plans to adopt a new mode of financing by issuing "tokens", which are digital tokens that enable anyone to invest directly in the country.

### 5.2.2. Patient data management

This section focuses on the importance of the healthcare sector on a national and international scale, as well as some thoughts on the prospects for its improvement through Blockchain technology.

### 5.2.2.1.  General context

The introduction of new information and communication technologies is one of the vectors for the development of healthcare in Tunisia, enabling better access to care and equality in the face of disease. To further this trend, Tunisia organizes an annual International Digital Health Forum[90] .

Digital healthcare has progressed through the development of connected objects, mobile applications, robotic surgery, the advent of IT, the exploitation of Big Data...

According to the WHO[91] "Digital health (e-health) covers the use of information and telecommunication technologies in the field of health and well-being." Thus, the three sectors of digital health are :

---

[89] "X-Road, available at https://e-estonia.com/solutions/interoperability-services/x-road/;

[90] "2nd International Digital Health Forum, available at http://africaine-sante.com.tn/les-grands-dossiers-dafricaine-sante/2eme-forum-international-de-la-sante-numerique/;

[91] Safon M., (2018), "La e-santé : Télésanté, santé numérique ou santé connectée", Centre de documentation de l'Irdes (Bibliographie thématique), page 4

- Tools and media: these include telematics (medical computing, information systems, medical records, etc.), the Internet and healthcare applications.
- Tele-services: corresponding to health information, wellness technologies, e-learning and commercial e-health services (tele-consulting, etc....).
- Regulated medical practices: essentially telemedicine (clinical)

Healthcare is a particularly promising sector for Blockchain technology. This technology could be used to develop numerous applications for the management, security and use of patient data. But regulatory and technical challenges remain significant. The various use cases are best considered in the medium and long term, given the complexity of integrating them into existing healthcare systems.

However, the pharmaceutical sector is more suited to the likely integration of Blockchain in the shorter term, to combat drug counterfeiting, and more generally improve the transparency and fluidity of its supply chain.

### 5.2.2.2. Blockchain enhancement

The fundamental advantages of a blockchain system lie in data integrity and networked immutability. For the healthcare industry, this concept can solve the emerging problems of the rapidly evolving, highly interconnected digital healthcare ecosystem. To achieve this, trust and governance will be critical implementation success factors for Blockchain-enabled applications and care management systems.

The aim is to modernize healthcare provision at all levels, and to provide citizens with simple, easy access to healthcare services through information systems, telemedicine, connected objects, etc.

Indeed, the use of new technologies such as Blockchain will enable better access to healthcare for all, and in particular for disadvantaged areas. To achieve this, we can identify three possible projects: shared computerized medical records, claims control and billing management, and traceability of the drug supply chain.

### Conclusion

The approach in this chapter was to make the correspondence between Blockchain technology and the modernization of the Tunisian administration. The interest was in identifying likely avenues for improving public services, focusing on evidence of the involvement of Tunisian public structures, such as the BCT and the Tunisian Post Office, in the process of improving their services through Blockchain technology, while trying to keep the citizen at the center of their concerns.

In addition, through the various projects proposed, we can identify two strategic axes on which the State can act. The first involves optimizing the efficiency and organization of the State. Indeed, in this approach, the industrialization and generalization of the various proposed solutions, such as that of the commercial register, to all registers can enable the dematerialization of the issuance of official documents and the disintermediation of their use. The second axis focuses on encouraging digital transformation, innovation, scientific research and economic development.

To achieve this, experimentation in a number of public and private sectors is the best way to develop usage. In the next chapter, we'll take a closer look at patient data management in the healthcare sector.

# Chapitre III. Blockchain and online services in Tunisia - the case of healthcare

## Introduction

Today, healthcare is taking an innovative approach to disease prevention and treatment that incorporates a patient's lifestyle and environment. This change can be delivered by Blockchain technology, which has the potential to be the technical standard that enables individuals, healthcare providers and entities, and medical researchers to share healthcare data in electronic format. Thus, in this chapter, we will describe the healthcare system in Tunisia (1), make an analysis of the Tunisian healthcare system environment (2) and end by proposing a Blockchain model dedicated to healthcare via a technical study (3) supported by some recommendations (4).

## 1. Tunisia's healthcare system

We can describe the health system in Tunisia by citing the current legal framework for the right to health in Tunisia (1.1) before describing the health information system (1.2) as well as defining the potential sectors to be affected by change using Blockchain technology (1.3).

### 1.1. Legal framework of the right to health

The legal and institutional framework encompassing the Tunisian healthcare system has developed significantly and is extremely complex and extensive. In addition, the enshrinement of the right to health in the new constitution of January 2014 constitutes an important step forward. Indeed, this can be revealed on three levels, illustrated by Table 3. Firstly, the constitutional value that concerns the health sector mentions its strategic importance for the State. Secondly, through Article 38 of the Constitution, a clear framework of the State's obligations towards its citizens in terms of health interventions has been established. Finally, we cannot speak of equity and universality of the right to health without having the right to information or without emphasizing the role of local authorities in terms of decentralization and local governance, and this is regulated through articles 139 and 140.

| Levels | Article |
|---|---|
| Constitutional | **Article 38**<br>Health is the right of every human being.<br>The State provides all citizens with preventive health care and provides the means necessary to guarantee the safety and quality of health services.<br>The State guarantees free health care for people without support and on low incomes. It guarantees the right to social coverage as provided by law. |
| Government bonds | **Article 139**<br>Local authorities adopt the instruments of participatory democracy and the principles of open governance to ensure the widest possible involvement of citizens and civil society in preparing development and land-use projects and monitoring their implementation, in accordance with the law. |
| Local governance | **Article 140**<br>Local authorities can cooperate and create partnerships with each other, with a view to carrying out programs or actions of common interest.<br>Local authorities can also establish external partnerships and decentralized cooperation.<br>The law defines the rules for cooperation and partnership. |

**Table 3Constitutional legal framework of the right to health**

## 1.2. Health Information System

Health services in Tunisia are divided geographically into three levels, provided by the public, parapublic and private sectors. The geographical distribution, illustrated in figure 9 below, is at first sight satisfactory. However, this geographical distribution of structures conceals an undeniable lack of availability and capacity to offer continuous and regular services.

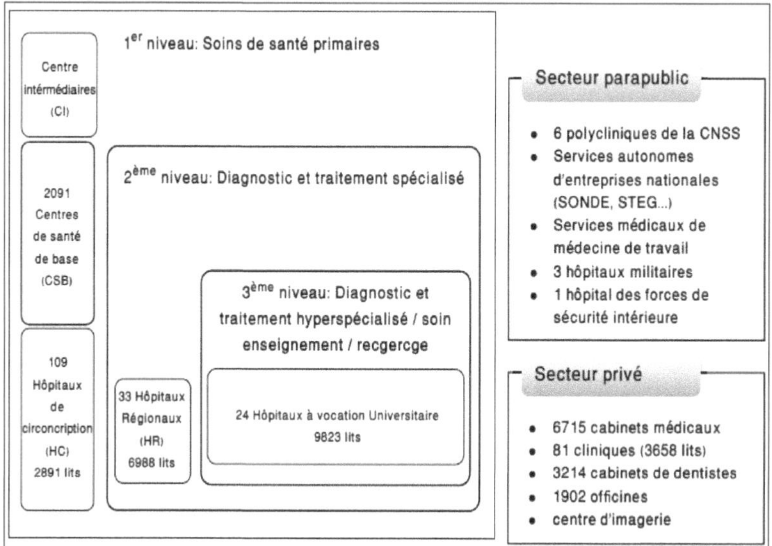



**Figure 8Geographical distribution of health services**[92]

On the other hand, the development of "Digital Health" in Tunisia is a strategic axis of the "Plan National Stratégique Tunisie Digitale 2020". Indeed, multiple factors, such as the complexity of the healthcare system, the multiplicity of the various stakeholders in the field, as well as the political, socio-sanitary and economic stakes of healthcare, contribute to the need to consider the modernization of the Health Information System (HIS) as an essential lever for any steering of reform in the healthcare sector.

The current SIS[93] does not provide a complete picture of reality on the ground. Indeed, the players using the SIS express their inability to obtain data, in sufficient quantities, reliably and in a timely manner. This has a negative impact on the decision-making process and on the way in which the quality of service provided to citizens is assessed.

---

[92] Ben salah F. Benhafaiedh A. , et al , (2014), "Santé en Tunisie, état des lieux", @ Comité technique du dialogue sociétal pour les politiques, stratégies et plans nationaux de santé (report), 194p.

[93] "Programme de Développement de la "Santé Numérique" en Tunisie", available at http://www.santetunisie.rns.tn/fr/prestations/programme-de-d%C3%A9veloppement-de-la-%C2%ABsant%C3%A9-num%C3%A9rique%C2%BB-en-tunisie ;

However, the SIS makes it possible to obtain a significant amount of information from the various administrative units reporting to the Ministry of Health (MoH) at all levels. But the problem lies in the lack of information and reliability of this data, which covers a wide range of subjects such as epidemiological data (mortality, morbidity, etc.), data relating to the social determinants of health and social and territorial inequalities, etc.

In addition, the HIS suffers from a multitude of challenges in terms of fragmentation, compartmentalization, lack of integration between numerous HIS subsystems, partial deployment of new technologies as well as an absence of the notion of data sharing with little focus on the patient and his or her care pathway.

Various SIS domains or components, such as the Hospital Information System (HIS), have not undergone any major changes since the reform carried out in the early 1990s. In addition, interconnection with IS belonging to other ministries for the sharing and collection of data related to the population's health status and the social determinants of health has remained minimal.

With the aim of remedying these various problems, the modernization of the SIS envisaged by the Ministry of Health has been validated by the National Strategic Plan Tunisie Digitale 2020 by allocating a provisional budget of around 70 million Dinars over the 2017-2020 period, enabling the financing of the conduct of a range of strategic projects including:

- The modernization of the Health Information System in Tunisia through the deployment of a network infrastructure to interconnect the various users of the Information System.
- The modernization of Hospital Information Systems to improve patient services and minimize operating costs. This is reflected in various projects such as the deployment of the Electronic Medical Record (DMI), the securing and control of the medication circuit from the Central Pharmacy to the Patient, the digitization of medical archives and the development of national registers for clinical and scientific research.

On the other hand, and to support the implementation of priority e-health projects such as the Dossier Médical Informatisé (DMI) and the digital medical archive.... A Technical Cooperation[94] was established in 2016 between the Agence Française du Développement (AFD), the Caisse Nationale d'Assurance Maladie (CNAM), the Ministry of Health (MS) and the Ministry of Social Affairs (MAS).

On the other hand, the local pharmaceutical industry[95] , which is an important component of Tunisia's health fabric, has undergone considerable development, thanks to the introduction of a legal and regulatory framework covering the manufacture, registration and marketing of medicines.

A multitude of administrative structures, such as the National Drug Control Laboratory, are involved in the management, analysis and control of pharmaceutical products and medicines, as well as the inspection of manufacturing units, in collaboration with the WHO.

Finally, seeing this diversity of interacting elements within the Tunisian public health ecosystem, it can be argued that interoperability is an essential element of any infrastructure supporting patient-centric outcomes research and the precision medicine initiative. Thus, a Blockchain-based national healthcare IT infrastructure has considerable potential to promote the development of precision medicine, advance medical research and invite patients to be more responsible for their health.

### 1.3. Potential areas

The areas that are likely to cause the most problems are access to healthcare data and services, diversity in the way healthcare records are processed, and lack of technical interoperability.

---

[94] "Programme de Développement de la "Santé Numérique" en Tunisie", available at http://www.santetunisie.rns.tn/fr/prestations/programme-de-d%C3%A9veloppement-de-la-%C2%ABsant%C3%A9-num%C3%A9rique%C2%BB-en-tunisie ;

[95] "Evolution of the local pharmaceutical industry", available at http://www.santetunisie.rns.tn/images/actualite-fr/dpm.htm;

The development of a healthcare IT system running on Blockchain technology can provide technological solutions to these and many other challenges, including healthcare data interoperability, integrity and security, portable user data and other areas.

More fundamentally, Blockchain technology could enable secure, irrevocable data exchange systems. This would enable seamless access to historical patient data while eliminating the clutter and cost of data reconciliation. Patient data can be seamlessly integrated into any system via a unique national identifier.

Thus, the deployment of a digital national identity with biometric data may prove useful in creating patient identifiers for the entire Tunisian population and providing secure login and session management. However, given the uncertainties surrounding data ownership and the governance structure for health data exchange between public and private entities, it would be difficult to replicate the Blockchain-secured health record model in a compliant manner.

In addition, a percentage, which can be estimated, of healthcare costs are fraudulent, resulting from exorbitant billing or billing for care not performed. A system based on Blockchain technology can provide effective solutions to minimize these medical billing frauds. Medical billings can therefore be made available through the individual's national identity. By having all citizens on the system, the billing process becomes easy.

The process can then be linked to the bank account, insurance claims, etc. By automating the majority of settlement and payment processing operations, the Blockchain system could eliminate data inconsistency and reduce administrative costs as well as service processing time for the various players in the healthcare ecosystem.

In addition, such a system could lead to interesting ramifications for improving some of the obstacles associated with tracking logistical information in order to develop reliability-centered maintenance functions.

In addition, a system based on Blockchain technology could foster collaboration between citizens and researchers around innovation in various fields such as medical research, precision medicine and population health management.

## 2. Analysis of the Blockchain environment in the Tunisian healthcare system

Strategic management in the healthcare sector is difficult. Indeed, the healthcare environment is subject to a great deal of uncertainty, linked to the various demographic, social and health, technological and economic pressures that are exerted.

Thus, making a strategic diagnosis of this sector, in terms of integrating Blockchain technology, actually involves using a number of models and analysis tools such as the PESTEL model (2.1) and the SWOT model (2.2).

### 2.1. PESTEL analysis

The development of a strategy for the deployment of new technologies and work techniques must necessarily be preceded by a phase of analysis and diagnosis from a number of angles.

The PESTEL method[96] is an analysis technique that uses a set of external variables or factors to anticipate the opportunities and threats linked to the environment of the structure or activity under study. In general, PESTEL analysis is the first step in developing a strategic analysis for a strategic business area (SBA).

This method lists six angles of macro-environmental influence that can have an impact on the activity or structure studied, which in our case is the healthcare sector:

---

[96] Besson B. et al , (2010), "Méthodes d'analyse appliquées a l'intelligence économique", @ ICOMTEC Poitiers (white paper), page 7

✓ **The political angle:** this concerns all decisions taken by governments on a national or international scale, such as fiscal policy, foreign cooperation, administrative modernization, social protection, etc.

✓ **The Economic angle:** factors influencing the purchasing power and behavior of customers or service consumers. These factors include inflation, unemployment, service costs, disposable income, interest rates, etc.

✓ **Sociological angle:** this angle deals with the social characteristics that influence the financial capacity of the service applicant. Examples include demographics, work attitude, social mobility, level of education, etc.

✓ **Technological angle:** this concerns all technological innovations that can disrupt existing supply and demand. This concerns, for example, private or public investment in R&D, patent development, speed and capacity for technology transfer....

✓ **The ecological angle:** this covers all the constraints associated with sustainable development in the healthcare sector, such as medical waste treatment, energy consumption, etc.

✓ **The Legal angle:** this angle covers all the regulations and legislation governing the business and the players in its ecosystem. This concerns the right to health coverage, personal data protection, health insurance, general health legislation, etc.

The tables below (Tables 4 to 9) illustrate our PESTEL analysis, listing the various factors, the key elements in relation to these factors, their influence, which can be positive (➚ ), negative (➘ ) or both (➚➘ ), and the degree of this influence, which varies from weak (◇ ) to medium (◈ ) and finally strong (◆ ), while marking an **x** in the appropriate column. The last column is used to identify questions formulated in relation to the influence factor. This analysis was based on the proposals submitted to the 2014 national health conference[97] .

---

[97] Ben salah F. Benhafaiedh A. , et al , (2014), "Pour une meilleur santé en Tunisie, Faisons le chemin ensemble", @ Comité technique du dialogue sociétal pour les politiques, stratégies et plans nationaux de santé (Livre blanc), 194 p.

| Factors Policies | Key elements | Influence | | | Degree | | | Related questions |
|---|---|---|---|---|---|---|---|---|
| | | ↗ | ↘ | ↗↘ | ◇ | ◈ | ◆ | |
| Regional policy | Clear legal basis, political decentralization, local management capacity, interaction between the health system and other sectors | x | | | x | | | Does a Blockchain-based system enable the adoption of a clear, up-to-date vision of a regional or even local healthcare policy that corresponds to the real specifics of the sector and regions? |
| Technological and health inequalities | Health coverage, health facilities, medical density, resources (infrastructure), personnel...) | | | x | | | x | Can we reduce social inequalities in healthcare and the major regional disparities in technological equipment? |
| Continuity of service | Discontinuity of treatment, discontinuity of supply, availability of service providers, coordination between the three levels of care, health check deadlines | x | | | | | x | Does the integration of Blockchain technology lead to a permanent organization and management of material and human resources? |
| Harmonization of practices relating to the use of patient medical records | Fragmentation of decisions, duplication of effort, collaboration Intersectoral, attitude of regional managers | x | | | | | x | Can we reduce the lack or discontinuity of coordination throughout the three levels of care? |

**Table 4PESTEL analysis - political factors**

| Factors Economic | Key elements | Influence | | | Degree | | | Related questions |
|---|---|---|---|---|---|---|---|---|
| | | ↗ | ↘ | ↗↘ | ◇ | ◈ | ◆ | |
| Optimizing healthcare facility expenditure | Budget, purchasing, maintenance, paper consumption, smooth relations between the administration, citizens and businesses | x | | | | x | | Does blockchain enable a gain in competitiveness and a reduction in costs? |

Table 5PESTEL analysis - economic factors

| Factors Social | Key elements | Influence | | | Degree | | | Related questions |
|---|---|---|---|---|---|---|---|---|
| | | ↗ | ↘ | ↗↘ | ◇ | ◈ | ◆ | |
| Image of the healthcare facility | Overall quality of service, patient/customer confidence, socio-economic indicators, quality standards | x | | | | x | | Is blockchain technology capable of democratizing access to services and promoting a more decentralized and horizontal relationship? |

Table 6PESTEL analysis - social factors

| Factors Technological | Key elements | Influence | | | Degree | | | Related questions |
|---|---|---|---|---|---|---|---|---|
| | | ↗ | ↘ | ↗↘ | ◇ | ◈ | ◆ | |
| Mitigation of cybercrime risks | Public and private key management, privacy protection, intrusion, web attacks | | | x | | x | | Is a blockchain-based system a panacea for IT security? |
| Integration of IT subsystems | Business applications, connectors, database | | | x | | x | | Can we guarantee the interoperability of health hardware and software infrastructure via a Blockchain platform? |

| Factors Technological | Key elements | Influence | | | Degree | | | Related questions |
|---|---|---|---|---|---|---|---|---|
| | | ↗ | ↘ | ↗↘ | ◇ | ◈ | ◆ | |
| HR skills | Skills, know-how, technical complexity | X | | | | X | | Can anyone master this technology? |

**Table 7PESTEL analysis - technological factors**

| Factors Ecological | Key elements | Influence | | | Degree | | | Related questions |
|---|---|---|---|---|---|---|---|---|
| | | ↗ | ↘ | ↗↘ | ◇ | ◈ | ◆ | |
| Impact of shared medical record operations on the environment | Waste management and treatment, communicable diseases, population hygiene levels | X | | | | X | | Does innovation through Blockchain enable the improvement of environmental health and the fight against communicable diseases? |
| Energy optimization | Electrical energy | X | | | | X | | How can we alleviate the enormous energy consumption of a Blockchain-based system? |

**Table 8PESTEL analysis - ecological factors**

| Factors Legal | Key elements | Influence | | | Degree | | | Related questions |
|---|---|---|---|---|---|---|---|---|
| | | ↗ | ↘ | ↗↘ | ◇ | ◈ | ◆ | |
| Managing incidents linked to the use of shared medical records | Legal framework, agreements, authorizations, service delivery structures, critical incidents | | | X | | | X | Can we reduce the complexity of existing appeal and complaint mechanisms in administrative and judicial circuits? |
| Patient data protection | Rights protection, control systems, safety standards | | | X | | | X | Does the use of Blockchain technology provide an effective legal |

| Factors<br>Legal | Key elements | Influence | | | Degree | | | Related questions |
| --- | --- | --- | --- | --- | --- | --- | --- | --- |
| | | ↗ | ↘ | ↗ ↘ | ◇ | ◈ | ◆ | |
| | | | | | | | | framework for the shared Medical Record? |

Table 9PESTEL analysis - legal factors

Blockchain technology probably represents the beginnings of the next stage in web development. This evolution is derived from the technology that enabled the rise of Bitcoins. The database, built on a blockchain, is decentralized, transparent and ultra-secure. With this approach, the data concerned is impossible to falsify.

Today, Blockchain technology is finding more and more outlets, not only in the private sector, especially for activities linked to finance, product traceability and the exchange of goods and services, but also in public administrations.

The application of such technology enables us to be truly at the cutting edge of innovation, with international visibility. Research into the potential of integrating blockchain into public services is part of the modernization of Tunisian public administration. The aim is to experiment with this technology, which is a high-security data structure that enables accounting and descriptive information to be archived and updated without intermediaries guaranteeing the authenticity and integrity of the information. This technology promises improved efficiency in information processing, lower processing costs, and irreversible security and logging of any action or process triggered.

### 2.2.SWOT Analysis

The principle of the SWOT analysis[98] (Strenght, Weakness, Opportunities and Threats) is to draw up an inventory of the situation based on positive and negative factors of internal origin (strengths/weaknesses) and external origin (opportunity/threat). Internal

---

[98] Besson B. et al , (2010), "Méthodes d'analyse appliquées a l'intelligence économique", @ ICOMTEC Poitiers (white paper), page 7

or external positioning depends on the Blockchain technology and its various components in opposition to the overall environment in which it is developing. Table 10 presents the results of this analysis.

Blockchain technology offers many advantages for information technology. Blockchain is based on open source software that facilitates faster and easier interoperability between systems already in use in Tunisia's various healthcare structures. Such a system can evolve efficiently to handle ever larger volumes of data and more users of Blockchain technology.

Open Source solutions also generate innovation in the applications market. Healthcare providers and individuals alike would benefit from the wide range of application choices, and could select options to suit their specific needs and requirements.

The system can be developed in each separate structure. For example, in each local authority, existing healthcare structures in the region can work together to build a repository that takes into account the demographic, health, social and economic specificities of the system's members, thus creating a territorial healthcare identity specific to their region. This improves the quality of service to citizens, and provides a map of healthcare needs (new treatments, drugs, etc.).

Blockchain technology also addresses the challenges of interoperability within the healthcare IT ecosystem. This will eliminate the need to develop complex APIs for point-to-point integration between different systems.

Blockchain technology would enable patients, the healthcare community and researchers to access a shared data source for timely, accurate and comprehensive patient health data. What's more, data structures are flexible, scalable and could support unforeseen data that will become available in the future.

Another advantage lies in Blockchain's distributed architecture, which would enable built-in fault tolerance and disaster recovery. This is because data is distributed across many servers in many different locations. There is no single point of failure, and a disaster is unlikely to impact all sites at once.

In addition, Blockchain technology works with standard algorithms and protocols for encrypting data. These technologies have been highly analyzed and accepted as secure, and are widely used across all industries and in many government agencies.

On the other hand, Blockchain technology will offer many benefits to medical researchers, healthcare providers, caregivers and individuals. The creation of a single storage location for all health data on a local or national scale would serve research as well as personalized medicine. Indeed, health researchers need extensive and comprehensive datasets to advance understanding of disease, accelerate biomedical discovery, speed up drug development and design individualized treatment plans based on patients' genetics, life cycle and environment.

The shared data environment provided by Blockchain technology would provide a wide range of data for longitudinal studies by including patients from different socio-economic backgrounds and geographical environments.

A healthcare Blockchain system would expand health data acquisition to include data from populations of people currently underserved by the medical community. This would enable individuals to be classified into sub-categories, distinguishing between those who respond well to a specific treatment and those who are more susceptible to a particular disease.

In addition, doctors' ability to obtain more frequent data such as blood pressure or blood sugar levels would improve individualized care through specialized treatment plans based on outcomes and treatment efficacy.

In addition, real-time access to data would improve the coordination of clinical care in emergency medical situations, and also enable researchers and public health resources to rapidly detect, isolate and direct changes in environmental conditions that have an impact on public health.

Finally, a healthcare Blockchain system would certainly foster the development of a new generation of "smart" applications for the benefit of medical research and personalized treatments. Also, healthcare provider and patient would be able to participate, through shared information, in a collaborative and informed discussion about the best treatment options based on research rather than intuition.

On the contrary, we can point to some limitations in the adoption of Blockchain technology for healthcare applications, and especially in the choice between a public Blockchain, which is open and accessible to all, or a private one, which is deployed within an institution or network. Indeed, the lack of data confidentiality and the absence of robust security make some public Blockchains unsuitable for a healthcare blockchain, which requires confidentiality and controlled, auditable access. In addition, scalability issues arise for large-scale, widely-used Blockchain applications.

Private blockchains, on the other hand, can address concerns about confidentiality, security and scalability. However, these Blockchains pose different challenges linked to the vendor-neutrality of solutions and the non-use of open standards.

| | Positive factors | Negative factors |
|---|---|---|
| **Internal diagnosis** | **FORCES** | **WEAKNESSES** |
| | <ul><li>Operational efficiency</li><li>Facilitates information sharing. No need to transmit multiple documents. Registration can be done directly on the Blockchain.</li><li>Secure encryption and tamper-proof data storage</li><li>Eliminates central authority with full access to data</li></ul> | <ul><li>Business rules change frequently, but this is not the case for Blockchain.</li><li>Blockchain is generally not modular. This means that an old encryption module cannot be easily replaced.</li><li>What happens if the business rules change and we want to export the data to a new blockchain with the correct data models? A blockchain doesn't provide an immediate, out-of-the-box exit strategy.</li><li>Blockchain technology potentially conflicts with existing approaches to regulatory compliance, and primarily personal data regulations.</li><li>The concept is not easy for everyone to grasp. Good education or training is needed to make mass adoption possible.</li></ul> |
| | **OPPORTUNITIES** | **THREATS** |

|  | **Positive factors** | **Negative factors** |
|---|---|---|
| **External diagnostics** | <ul><li>Provides a platform for Big Data and analytical research.</li><li>Returns control to the user, for example, everyone can control who has access to this data, and all these authorizations will be stored on the Blockchain.</li><li>The emergence of digital through the national or local healthcare network means that more people will accept the concept of Blockchain in their daily lives.</li></ul> | <ul><li>Scalability problems: too many transactions lead to overload, even though several solutions are available.</li><li>A fast-changing environment</li><li>Attacks and hacking are always a possibility.</li></ul> |

**Table 10:SWOT analysis of Blockchain technology dedicated to the healthcare sector**

## 3. Technical study

Health data may not be available in an identical form. As a result, the various players in the healthcare ecosystem, such as patients, public authorities, the medical profession and researchers, cannot access it to offer personalized diagnosis and care. Added to this problem is the issue of interoperability between the systems and technologies used. This makes it all the more interesting to propose optimal digital solutions for healthcare.

This technical study clarifies the solution adopted to overcome these challenges. It covers the specification of the requirements (3.1) and a technical description of the chosen Blockchain model (3.2).

### 3.1.Specifications

The specification includes a technical diagnosis (3.1.1), a requirements specification (3.1.2) and a delimitation of the system modules (3.1.3).

### 3.1.1. Diagnosis technical

Using the report drawn up in December 2014 during the societal dialogue for national health policies, strategies and plans[99] as well as an inventory of the Tunisian healthcare system focused on the public sector presented at the second International Digital Health Forum held in February 2017[100] , the following conclusions can be drawn:

❖ **Infrastructure**

There is insufficient deployment of the national healthcare network: the digitization of health information systems (HIS) and hospital information systems (HIS) requires first and foremost a high-quality network infrastructure to enable the exchange of information for the care of citizens.

❖ **Information systems**

The diagnosis showed, firstly, that there is a lack of standardized medical procedures and virtually no culture of sharing. Indeed, certain medical procedures require standardization and unification.

Secondly, the Information System is not currently focused on the patient and his or her care pathway. This is due, on the one hand, to the lack of a legal reference framework governing the Medical Record; and, on the other hand, to the fact that the Medical Record does not, in some configurations, enable a patient's care path to be traced, given the multiplication of medical records created by each department, the lack of

---

[99] Ben salah F. Benhafaiedh A. , et al , (2014), "Santé en Tunisie, état des lieux", @ Comité technique du dialogue sociétal pour les politiques, stratégies et plans nationaux de santé (report), 194p.

[100] Helmi I., (2017), "e-santé Tunisie -Vers un système de santé connecté", @ 2ème session of the International Digital Health Forum (conference), Hammamet, 31 p.,

integration of SISs from different health establishments, and the absence of a Unique Identifier.

These shortcomings generate avoidable additional costs, undermine the quality of patient care and hinder the exchange of experience between healthcare professionals.

Thirdly, there is a very partial deployment of digital solutions in the SIS space, due to a lack of integration between all the players in the healthcare system, such as payers (CNAM, insurers, etc.), the central pharmacy, the private sector (outpatient medicine, laboratories, clinics, etc.) and the other players in the healthcare ecosystem.

❖ **Medicines**

There are gaps in the end-to-end traceability of the drug circuit, as well as slowness in marketing authorization procedures for new drugs and renewal applications.

### 3.1.2. Requirements specification

Improved patient flow management enables efficient follow-up care, and better utilization of services through better knowledge of the patient's actual condition via a medical record that is constantly updated in line with treatments and clinical diagnoses. This enables us to provide up-to-date information to the patient's family, ensuring overall satisfaction, as well as flexibility in billing for services rendered.

In addition, drug administration must ensure that the right dose of the right drug is administered to the right patient, at the right time, in the right order of care. In order to limit medical errors and their harmful consequences for patients and hospitals, who pay for them in financial compensation, it is important to take a holistic approach to patient care, which means building a patient-centric information system that takes into account from the outset how to integrate the multiple electronic systems already in use.

Indeed, healthcare professionals are missing three very important factors in the process of administering drugs and patients[101] . The first factor is time. Nurses and nursing managers are obliged to devote a percentage of their time to administrative tasks. The hospital pharmacist is totally absorbed by the administration of documents accompanying drug delivery. The order picker has to write down the doses of his drugs by hand. The OR manager has to manage implantable device stocks by hand. Added to these constraints are the working conditions. In other words, the conditions under which care activities are carried out no longer comply with the most elementary safety rules, and inevitably lead to incidents with known consequences.

The second factor is security, which is one of the elements imposed by the health authorities. However, all the procedures imposed represent a considerable volume of information, requiring technological resources that caregivers do not have.

The third factor is cost control, which has become the main preoccupation of plant managers. The sources of financial loss are too numerous to list exhaustively.

This is why an approach using Blockchain technology can significantly improve the quality of care, and thus provide tangible benefits for patients. The idea is to design an intelligent Blockchain platform for healthcare. In effect, this is a web and mobile application that enables the digital management of patient records by integrating meaningful user interfaces for patients and service providers to manage the information entered. Our application, illustrated in Figure 10, should enable:

✓ Management of a simple, secure web interface to automate the flow of information within a hospital or other healthcare facility.
✓ The speed of operations for drug traffic.
✓ The economics of pharmaceutical consumption.
✓ Secure access for citizens to their health data, and the sharing of this data with the various stakeholders in the health sector.

---

[101] Rabaoui A. and Messeoudi H., (2009), "Conception et réalisation d'une chaine logistique intelligente pour l'industrie pharmaceutique par RFID", @ ENISO (Mémoire de Projet de Fin d'Études), 65 p.

✓ Improving data quality to enhance medical research, disease prevention and personalized healthcare.
✓ Improved patient management by physicians.
✓  Standardization of computerized patient health records across Tunisia.
✓ Improving interoperability between healthcare systems and applications.

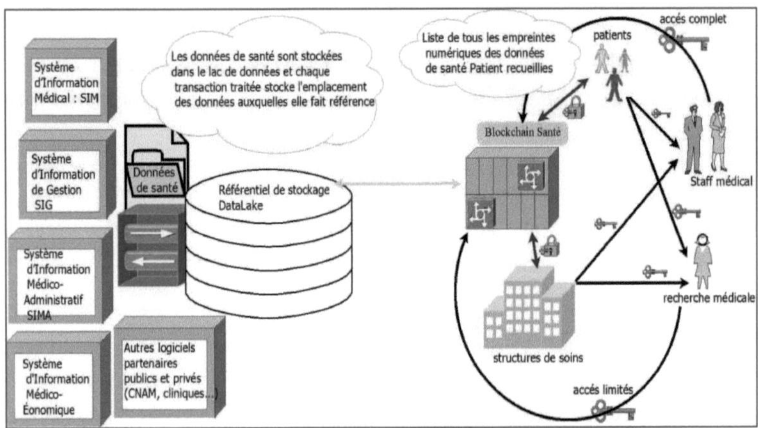


**Figure 9:Architecture of the proposed Health Blockchain application**

### 3.1.3. System module delimitation

We can identify three modules that can be integrated into the envisaged Blockchain platform: a module concerning the shared medical record (3.1.3.1), another for billing and claims management (3.1.3.2) and a final one *for* drug traceability (3.1.3.3).

#### 3.1.3.1.  Shared medical record module

The insurance file is a component of the Ministry of Health's DMI computerized medical file. The CNAM is modernizing its information system by integrating new technologies. It hopes to achieve better quality, better control and better management of the ceiling, as well as automatic processing of benefits. The aim is to guarantee access to care and better management of available resources.

To this end, digitization of the health insurance file is one of the components of the CNAM information system. The fund's strategic priorities include[102] :

- The dematerialization of care sheets through the definition of a care basket for each APCI and a therapeutic protocol (defining the most appropriate drugs) and another for reimbursement;
- Dematerialization of requests for services subject to prior agreement ;
- The introduction of an electronic format for hospital discharge documents.

In line with the CNAM's strategic axes, Blockchain technology can be introduced to create a shared computerized medical record. This digital file enables authorized healthcare professionals to access information useful for the care of an insured person, and to share this medical information with other healthcare professionals: medication, hospitalization and consultation reports, examination results (X-rays, biological analyses, etc.), etc.

It should be similar to a health record, always secure and accessible to authorized healthcare professionals (doctors, nurses, pharmacists, etc.).

In Estonia, for example, Blockchain company Guardtime and the eHealth Foundation announced a partnership in March 2016[103] , with the deployment of a Blockchain-based system to secure over a million medical records.

The challenge is that digital health revolves around patients' personal data. In fact, health data is highly sensitive; so, to make progress on a computerized medical record project, you need a legal framework that protects personal data.

---

[102] Irmani B. et al , (2017), "Le Système d'Echange Electronique des Données : Le volet assurance maladie du Dossier Médical Informatisé",@ 2ᵉᵐᵉ session of the International Digital Health Forum (conference), Hammamet, 39 p.;

[103] "The Blockchain Decrypted," available at https://managersante.com/2017/11/27/revolution-blockchain-sante-adnanelbakri/;

In conclusion, the modernization of healthcare services is both a strategic choice and an obligation, to ensure the continuity of healthcare services throughout Tunisia and to establish good governance of the various health structures.

### 3.1.3.2. Billing management module and claims management

Blockchain-based systems can provide realistic solutions for minimizing medical billing fraud. These frauds result from false billing or excessive billing for services not performed.

What's more, these systems could help reduce administrative costs and delays for suppliers and payers.

### 3.1.3.3. Drug traceability module

To combat counterfeit drugs, a system based on Blockchain technology could provide logging and tracking of every step in the supply chain at the pharmaceutical product level. In addition, using private keys and smart contracts, it would be possible, on the one hand, to prove ownership of the drug source at any point in the supply chain and, on the other, to manage contracts between the various parties.

An example of this is a supply chain use case developed by IBM[104] . The solution monitors an entire process in an international distribution chain. Each participant can intervene and control the stage for which he or she is responsible.

### 3.2. Technical description of the Blockchain model for shared patient records

In this section, a description of the chosen Blockchain model is given, clarifying how it works (3.2.1), as well as data access and confidentiality security guarantees (3.2.2).

### 3.2.1. How it works

---

[104] "Blockchain, multiple use cases without cryptocurrency", available at https://thd.tn/la-blockchain-des-cas-dusages-multiples-sans-cryptomonnaie/ ;

In order to avoid any risk associated with storing public data on the Blockchain, it is possible to design a solution that enables not a piece of data, but its digital fingerprint to be stored via Blockchain technology, in such a way that the data remains stored off-Blockchain.

Our proposal involves the use of a Blockchain-based architecture as an access control manager for computerized patient health records.

Any Blockchain for healthcare, and in particular for the shared patient record, should also include technological solutions for three key elements which are scalability, access security and data confidentiality.

A distributed blockchain containing health records, documents or images would have implications for data storage and data throughput limitation. Indeed, since health data is dynamic and expansive, replicating all health records on every member of the network would require a lot of bandwidth, waste network resources and pose data throughput problems.

For healthcare to take advantage of Blockchain technology, it is useful to develop a framework that should function as an access control manager for healthcare records and data.

The information contained in the proposed Health Blockchain framework would be similar to an index or catalog that contains a list of all the patient's health records. Transactions within blocks would contain a user's unique identifier, an encrypted identifier linked to the health data record, and a timestamp indicating when the transaction was created.

To improve the efficiency of data access, the transaction would contain the type of data contained in the health record and any other metadata that would facilitate frequently used queries. In this way, the Health Blockchain would contain a complete indexed history of all medical data, including formal medical records as well as health data from the likely use of mobile apps. This data would follow an individual user throughout his or her life.

In the first stage, all medical data would be stored outside the Blockchain in a data repository called a data lake[105] . A data lake is a storage repository that allows a volume of raw data to be kept in its native

---

[105] Tomcy J. and Pankaj M. , (2017), "Data Lake for Enterprises - Leveraging Lambda Architecture for building Enterprise Data Lake," Copyright © 2017 Packt Publishing, pages 38 - 43

format until it is useful for exploitation. This data lake would make healthcare research and analysis easy. Factors leading to specific outcomes could be explored, optimal treatment options determined and preventive medicine improved. What's more, all the information stored in the data lake would be encrypted and digitally signed to ensure confidentiality and authenticity.

Conceptually, the following figure (figure 11) summarizes the use case for this data repository. Structured, semi-structured and unstructured data are fed into the Data Lake, from which a Single Customer View (SCV) is holistically derived.

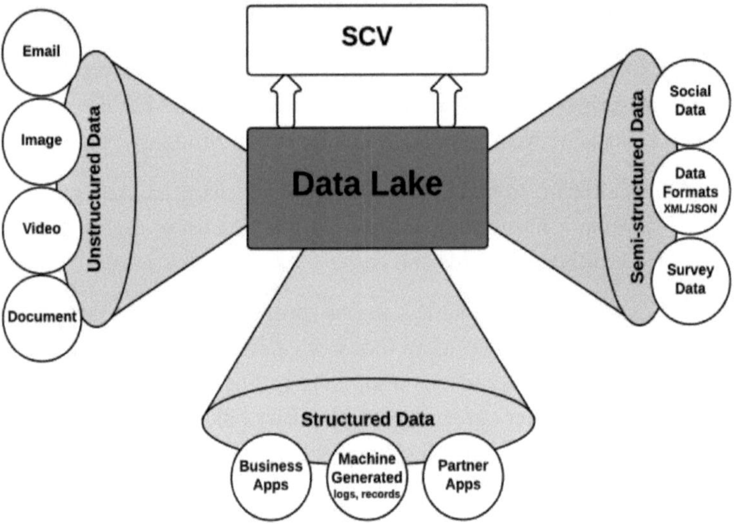


**Figure 10Conceptual view of the Data Lake use case for a SCV[106]**

In a second stage, the information stored in the data lake will be directly integrated into a separate blockchain accessible to all members of the system.

When a healthcare actor, such as a doctor or radiology technician, etc., creates a medical record (prescription, lab test, MRI...), a digital signature is created to verify the authenticity of the document or image.

[106] Tomcy J. and Pankaj M. , (2017), "Data Lake for Enterprises - Leveraging Lambda Architecture for building Enterprise Data Lake," Copyright © 2017 Packt Publishing, page 36

Health data would be encrypted and sent to the data lake for storage. Each time information is recorded in the Data Lake, a referent to the health record is added to the Blockchain in association with the user's unique identifier. A notification will be sent to the patient indicating that health data has been added to his or her Blockchain.

In a third stage, and as the system matures, a patient will be able to add health data using digital signatures and encryption from mobile applications and wearable sensors made available to him or her.

### 3.2.2. Access security and data confidentiality

The user would have full access to their data and control over how their data is shared, but without the ability to delete data. The user will assign a set of access permissions and designate who can query and write data to their Blockchain.

A mobile or desktop application would allow the user to see who has permission to access his blockchain. The user could also view an audit log showing who has accessed their blockchain, including when and what data they have accessed. The same dashboard would allow the user to give and revoke access permissions to anyone with a unique identifier. This provides an environment of transparency and allows the user to make all the decisions about what data is collected and how it can be shared.

Ideally, it is desirable to use a biometric identity system, since it offers improved security over password- and smart card-based methods of identity authentication.

Deploying such a Blockchain-based system enables better collaboration between the patient, specific healthcare providers and various healthcare entities. The decentralized nature of Blockchain, combined with digitally signed transactions, ensures equality of healthcare entitlements.

What's more, storing data in a separate repository protects it from illegal intrusion into the shared registry, since only hashed pointers and encrypted information are contained in the transactions.

This system can also be used to manage billing. Indeed, a block is created in the Blockchain network when peer nodes of a set of validators execute a consensus on a set of transaction results according to a verification logic that specifies under which conditions a transaction can be executed and whether the contractual conditions are satisfactory.

Taking health insurance as an example, we consider a simple scenario where the main processes are standard insurance transactions such as customer registration, claim submission, refund processing, etc... The Blockchain maintains the execution and results of each transaction and ensures that customers do not provide false statements to the insurance company, and that the insurance company is accountable for all the services it provides.

## 4. Recommendations and the way forward

Innovative digital solutions based on Blockchain technology can improve the organization of healthcare services and provide quality care for Tunisian patients. To do this, we need to map out a way forward, including a list of recommendations drawn from the experiences of governments and public institutions around the world that have tested this technology.

With the help of documentary resources available on the Internet, we were able to consult a number of English- and French-language books and articles dealing with this topic, enabling us to identify, for Tunisia, the path to follow for an effective and efficient deployment of Blockchain technology within the healthcare sector.

Firstly, a technology-based solution must be designed to meet the needs of citizens and existing healthcare systems. Its implementation must be thoughtfully tailored to the local, regional and national context.

Secondly, the adoption of such a solution must be seen as an integral part of healthcare and its IT systems, so as to support the measures and strategies adopted by the Tunisian government to reform healthcare systems.

What's more, the active commitment of all stakeholders in the healthcare ecosystem is needed to achieve a result that is useful to both citizens and public health authorities. It is by working together that we can achieve the rapid deployment of innovative digital solutions in healthcare and other fields, with minimal impact.

Also, attention must be paid to improving existing legislation on personal data protection, electronic identification and information security by introducing measures linked to the use of blockchain technology.

Finally, defining specifications for a unified format for the exchange of electronic health records is appreciated in the interests of promoting sustainable development and preventing the potential exploitation of data for research and other purposes.

All these measures could also lead to the development of more robust regional healthcare ecosystems.

**Conclusion**

The vision set out in this chapter is to promote healthcare, by offering a solution based on Blockchain technology that will help meet patients' unmet needs and facilitate equal access to high-quality care. This solution will also strengthen the resilience of Tunisia's health and care systems and make them more sustainable.

We decided to focus on healthcare because it's a complex sector that affects the entire Tunisian population. Exploring Blockchain technology could have a relevant impact on several processes and application scenarios. Consequently, use cases in this sector could help identify the advantages and disadvantages of the technology itself.

A platform using Blockchain technology could enhance trust, traceability and transparency. Blockchain is thus referred to as a vital resource in trusted computing.

# General conclusion

Data is an important part of the modern world's infrastructure, as it is essential to the functioning of society and a key factor in any digital transformation. For this reason, we need to be careful about how we build, maintain and strengthen our data infrastructure. Blockchain is a potentially important technology that enables a shared data infrastructure, and one that deserves to be investigated. Hence the interest of this ENA graduate thesis.

However, the emergence of new technologies, such as Blockchain, always goes through a pre-integration cycle. The challenge at the start of this cycle is to identify uses and applications that will stand the test of time, especially when handling personal data.

Blockchain technology could be used to strengthen trust in government services through public verifiability. It also offers great potential for collaborative data maintenance, as well as the collection and publication of widely distributed data for applications dealing with various fields.

Many ideas have been mooted in this report that would introduce new methods of data processing via Blockchain technology, but if deployed incorrectly, this would create significant new privacy issues. Thus, the successful design of the data infrastructure will come from involving the political and technical decision-makers in each relevant sector, identifying common challenges and determining the most appropriate technological approaches to solving them.

As a result, the healthcare sector was chosen for an in-depth study into the integration of Blockchain technology. The choice of this sector is explained by the fact that the adoption of digital solutions in this field remains slow, and varies greatly from region to region, in addition to the interoperability issues that existed in the hardware and software infrastructure.

The most effective approach to advancing interoperability objectives within the healthcare sector would be to establish a national technology infrastructure based on open standards. In addition, a shared distributed infrastructure that provides a complete view of a patient's health data throughout his or her life is an equally essential element for the smooth operation of interoperable healthcare IT systems.

In this way, Blockchain technology meets the challenges of interoperability, is based on open standards, provides a shared view of healthcare data and will be widely accepted and deployed across all industries.

Indeed, through Blockchain technology, a decentralized database without intermediaries will be created. This technology enables the automation of a transaction, its authentication, the certification of its date and the identity of the parties involved, the storage of the transaction data in a perennial and unalterable way and the guarantee of the quasi-inviolability of this data while ensuring free access by interested parties.

Use of the proposed solution could enable Tunisian citizens, healthcare providers and medical researchers to share vast quantities of health data. The acquisition, storage and sharing of this data would provide a scientific basis for the advancement of medical research and precision medicine, and help identify and develop new ways of treating and preventing disease.

Blockchain technology could strongly wrest its place in the healthcare IT ecosystem, and Tunisia should strongly consider basing its interoperability strategy on this technology and its use to promote the advancement of precision medicine.

Beyond the technical aspects of Blockchain technology, it will facilitate the implementation of a vision of digital democracy for Tunisia. Indeed, this tool could contribute to an equitable evolution of society through the creation of new economic opportunities, in addition to a thorough and approved decentralization of public decision-making and a better inclusion of Tunisian citizens in the exercise of their fundamental rights.

As a result, no matter where they are in Tunisia, citizens will be able to benefit from secure access to a complete electronic file containing a wealth of data concerning their health. They would also be able to retain control over their health data, and share it securely with authorized third parties for various uses, regardless of where the data is located, in full compliance with data protection legislation.

By way of conclusion, we would like to point out that this current study is only the first step in a research process that needs to be taken further, in order to further target the objectives. In addition, it is necessary to progressively broaden the areas of exchange concerning Blockchain technology to cover the interoperability of the computerized healthcare systems of the various stakeholders, on a local, regional and national level, by supporting the development and adoption of a unified format for the exchange of computerized patient records. Last but not least, political support and flexible, participative leadership will be essential to effectively harness the leverage of dematerialization and lower intervention costs for administrations dealing directly with patients and citizens in general.

# Bibliography


1. AIMF, (2004), "Fonctionnement de l'état civil dans le monde francophone", @ AIMF, 39 p.

2. Anthes G., (2015), "Estonia: A model for e-government", Article in: Communications of the ACM (vol 58), N°6, 3 p.

3. Ben salah F. Benhafaiedh A. , et al , (2014), "Santé en Tunisie, état des lieux", @ Comité technique du dialogue sociétal pour les politiques, stratégies et plans nationaux de santé (report), 194p.

4. Ben salah F. Benhafaiedh A. , et al , (2014), "Pour une meilleur santé en Tunisie, Faisons le chemin ensemble", @ Comité technique du dialogue sociétal pour les politiques, stratégies et plans nationaux de santé (Livre blanc), 194 p.

5. Besson B. et al , (2010), "Méthodes d'analyse appliquées à l'intelligence économique", @ ICOMTEC Poitiers (white paper), 127 p.

6. Berbain C., (2017), "La blockchain: concept, technologies, acteurs et usages", Série trimestrielle -Réalités Industrielles - Blockchains and smart contracts: technologies of trust? , 128 p.

7. Blockchain France, (2016), "La Blockchain décryptée-Les clefs d'une révolution", Observatoire Netexplo © Blockchain France Associés, 130 p.

8. Brown D., (2005), "E-government and public administration", Revue Internationale des Sciences Administratives (Vol. 71), @ I.I.S.A., 192 p.

9. Chakchouk M., (2017), "Blockchain in Tunisia: From Experimentations to a Challenging Commercial Launch", ITU Workshop on "Security Aspects of Blockchain", Geneva, Switzerland, 9 p.

10. Delahaye J., (2016), "Monnaies cryptographiques & blockchains", Université de Lille 1@ INRIA Saclay, 169 p.

11. Digital Exploration, (2015), "Estonie se reconstruire par le numérique", (cc) creative commons - Renaissance Numérique, 22 p.

12. Ethereum community, (2017), "Ethereum Homestead Documentation, Release 0.1", Ethereum Homestead Documentation, 121 pp.

13. EVRY, (2016), "Blockchain: Powering the Internet of Value", Whitepaper @ evry's labs, 50 p. ; available at https://www.evry.com/en/news/articles/banking-on-the-blockchain/

14. Faridah D., and Gallouj F, (2012) "Innovation in public services", Revue française d'économie, volume xxvii, no. 2, pp. 97-142 ; Available at https://www.cairn.info/revue-francaise-d-economie-2012-2-page-97.htm

15. Gauche K. and Taddei R., (2011), "Enjeux et services de l'administration électronique locale. Etude de cas", Nouveaux usages de l'internet dans les collectivités territoriales, IAE NICE, France. 19 p.

16. Guillaume B., (2016), "Understanding Blockchain - anticipating Blockchain's disruptive potential on organizations", uchange.co, 55 p.

17. Helmi I., (2017), "e-santé Tunisie -Vers un système de santé connecté", @ $2^{ème}$ session of the International Digital Health Forum (conference), Hammamet, 31 p.

18. Henman P., (2010), "Governing Electronically: E-Government and the Reconfiguration of Public Administration, Policy and Power", First published by Palgrave Macmillan, 280 p.

19. INNORPI ; " Manuel d'utilisation du Site marchand du Registre Central du Commerce ", @ INNORPI /Manuel utilisation version 1, 26 p.

20. Irmani B. et al , (2017), "Le Système d'Echange Electronique des Données : Le volet assurance maladie du Dossier Médical Informatisé",@ $2^{ème}$ session of the International Digital Health Forum (conference), Hammamet, 39 p.

21. Labrot É. and Ségur P., (2011), "Un monde sous surveillance?", © Presses universitaires de Perpignan, 255 p.

22. La Poste Tunisienne, (2016), "Annuaire Statistique 2016", @Poste_Tn, 39 p.

23. Laurence T., (2017), "Blockchain for Dummies", John Wiley & Sons, Inc, Hoboken, New Jersey, 214 pp.

24. Magnen J. and Fourel C., (2015), "Mission d'étude sur les monnaies locales complémentaires et les systèmes d'échange locaux", report, Part One, sec.secacess-presse@cabinets.finances.gouv.fr, 76 p.

25. MEDEF, (2016), "La blockchain pour les entreprises - Soyez curieux! Understand and experiment", 60 p.

26. Morabito V., (2017), "Business Innovation Through Blockchain, The B³ Perspective", © Springer International Publishing, 173 pp.

27. Mishra R. et al, (2018), "How Integrated Process Management Completes the Blockchain Jigsaw", Digital Systems & Technology @Cognizant, 16 pp.

28. Nakamoto S., (2008), "Bitcoin: A Peer-to-Peer Electronic Cash System", 9 p.; available at www.bitcoin.org

29. Oberdorff H., (2006), "L'administration électronique ou l'e-administration", In: Recherches et Prévisions, n°86, La nouvelle administration. L'information numérique au service du citoyen, 109 p.

30. Tallinn Entrepreneurship Office, (2017), "Tallinn in brief 2017", Tallinn - economic center of Estonia, 56 p.

31. Rabaoui A. and Messeoudi H., (2009), "Conception et réalisation d'une chaine logistique intelligente pour l'industrie pharmaceutique par RFID", @ ENISO (Mémoire de Projet de Fin d'Études), 65 p.

32. Safon M., (2018), "La e-santé : Télésanté, santé numérique ou santé connectée", Centre de documentation de l'Irdes (Bibliographie thématique), 342 p.

33. Smart Dubai Office, (2017), "Dubai -the first city on the Blockchain", case study© Smart Dubai, 19 p.

34. Tapscott D. and Tapscott A., (2016), "Blockchain_Revolution", Penguin Random House LLC (version 1), 318 pp.

35. Tomcy J. and Pankaj M., (2017), "Data Lake for Enterprises - Leveraging Lambda Architecture for building Enterprise Data Lake", Copyright © 2017 Packt Publishing, 585 p.

36. UK Government Chief Scientific Adviser, (2016), "Distributed Ledger Technology: beyond block chain", Open Government Licence © Crown copyright, 87 p.

37. United Nations Department of Economic and Social Affairs, (2014), "United Nations E-Government Survey 2014", Copyright © United Nations, 284 p.

38. United Nations Department of Economic and Social Affairs, (2018), "United Nations E-Government Survey 2018", Copyright © United Nations, 300 p.